AF147831

Current Topics in Microbiology

244 and Immunology

Editors

R.W. Compans, Atlanta/Georgia
M. Cooper, Birmingham/Alabama
J.M. Hogle, Boston/Massachusetts · Y. Ito, Kyoto
H. Koprowski, Philadelphia/Pennsylvania · F. Melchers, Basel
M. Oldstone, La Jolla/California · S. Olsnes, Oslo
M. Potter, Bethesda/Maryland · H. Saedler, Cologne
P.K. Vogt, La Jolla/California · H. Wagner, Munich

Springer-Verlag Berlin Heidelberg GmbH

Immunoreceptor Tyrosine-based Inhibition Motifs

Edited by M. Daëron and E. Vivier

With 20 Figures and 5 Tables

 Springer

MARC DAËRON, M.D. Ph.D.
Laboratoire d'Immunologie
Cellulaire et Clinique
INSERM U.255
Institut Curie
26 rue d'Ulm
F-75005 Paris
FRANCE

ERIC VIVIER, DUM, Ph.D.
Centre d'Immunologie INSERM-CNRS de Marseille Luminy
Institut Universitaire de France
Parc Scientifique de Luminy – Case 906
F-13288 Marseille Cedex 9
FRANCE

Cover Illustration: Schematic representation of ITIM-bearing inhibitory molecules and corresponding activating molecules.

Cover Design: design & production GmbH, Heidelberg

ISBN 978-3-540-65789-7   ISBN 978-3-642-58537-1 (eBook)
DOI 10.1007/978-3-642-58537-1

This work is subject to copyright. All rights are reserved, whether the whole or part of the material is concerned, specifically the rights of translation, reprinting, reuse of illustrations, recitation, broadcasting, reproduction on microfilm or in any other way, and storage in data banks. Duplication of this publication or parts thereof is permitted only under the provisions of the German Copyright Law of September 9, 1965, in its current version, and permission for use must always be obtained from Springer-Verlag. Violations are liable for prosecution under the German Copyright Law.

© Springer-Verlag Berlin Heidelberg 1999
Originally published by Springer-Verlag Berlin Heidelberg New York in 1999

Library of Congress Catalog Card Number 15-12910

The use of general descriptive names, registered names, trademarks, etc. in this publication does not imply, even in the absence of a specific statement, that such names are exempt from the relevant protective laws and regulations and therefore free for general use.

Product liability: The publishers cannot guarantee the accuracy of any information about dosage and application contained in this book. In every individual case the user must check such information by consulting other relevant literature.

Typesetting: Scientific Publishing Services (P) Ltd, Madras

Production Editor: Angélique Gcouta

SPIN: 10714847   27/3020 – 5 4 3 2 1 0 – Printed on acid-free paper

List of Contents

List of Contributors

(Their addresses can be found at the beginning of their respective chapters.)

ALLEY, T.L.   137

BIASSONI, R.   69

BORGES, L.   123

BOTTINO, C.   69

BRAUD, V.M.   85

BURROWS, P.D.   137

CAMBIAGGI, A.   169

CAMBIER, J.C.   43

CHEN, C.C.   137

COLONNA, M.   115

COOPER, M.D.   137

CORNALL, R.J.   57

COSMAN, D.   123

CYSTER, J.G.   57

DAËRON, M.   1, 13

DAMEN, J.E.   29

DURONIO, V.   29

FANGER, N.   123

FONG, D.   43

FRIDMAN, W.H.   13

GOODNOW, C.C.   57

HELGASON, C.D.   29

HO, L.H.   137

HUBER, M.   29

HUMPHRIES, R.K.   29

HUREZ, V.   137

KRYSTAL, G.   29

KUBAGAWA, H.   137

LAM, V.   29

LÓPEZ-BOTET, M.   115

LUCAS, M.   169

MALBEC, O.   13

MCMICHAEL, A.J.   85

MEYAARD, L.   151

MILLO, R.   69

MORETTA, A.   69

NAVARRO, F.   115

PECHT, I.   159

SALCEDO, M.   97

SCHEID, M.P.   29

SHIMADA, T.   137

TAMIR, I.   43

TUN, T.   137

UEHARA, T.   137

VÉLY, F.   169

VIVIER, E.   1, 169

WAGTMANN, N.   107

XU, R.   159

Biology of Immunoreceptor Tyrosine-based Inhibition Motif-Bearing Molecules

M. Daëron[1] and E. Vivier[2]

1 Introduction

The term "immunoreceptor tyrosine-based inhibition motif" (ITIM) was coined after the term "immunoreceptor tyrosine-based activation motif" (ITAM) to designate molecular motifs that antagonize ITAM-dependent cell activation (Daëron et al. 1995a; D'Ambrosio et al. 1995). The concept of ITIM lies on the coexpression by single cells of "on" and "off" molecules which, when kept in close proximity, transduce intracellular signals that interfere with each other. On and off receptors can be constitutively associated at the cell membrane, such as B Cell receptors (BCR) and CD22 on B lymphocytes (Law et al. 1996) or high-affinity IgE receptors (FcεRI) and mast cell-associated function antigen (MAFA) on mast cells (Guthmann et al. 1995); or they can be coaggregated by extracellular ligands. Thus, ITIM-bearing low-affinity receptors for IgG (FcγRIIB) and ITAM-bearing receptors are coaggregated by IgG antibodies or soluble immune complexes (Amigorena et al. 1992; Daëron et al. 1995b; Daëron et al. 1995a). Although the spatial relationship established between MHC class I-specific killer cell Ig-like receptors (KIRs) and receptors that trigger cytotoxicity (Moretta et al. 1997) is unclear, KIRs needed to be coligated with FcεRI by the same soluble molecules in order to inhibit the activation of mast cells in a reconstitution model (Bléry et al.

[1] Laboratoire d'Immunologie Cellulaire et Clinique, INSERM U.255, Institut Curie, Paris, France
[2] Centre d'Immunologie INSERM-CNRS de Marseille-Luminy, Institut Universitaire de France, Marseille, France

1997). Following aggregation or coaggregation of the receptors, both ITAMs and ITIMs are tyrosyl phosphorylated by src family protein tyrosine kinases and, once phosphorylated, they recruit cytoplasmic molecules having SH2 domains (CAMBIER 1995; BURSHTYN and LONG 1997; VIVIER and DAËRON 1997). The antagonistic effect of ITIM-bearing molecules on ITAM-bearing molecules is based on the specificity of interactions of phosphorylated ITAMs with the SH2 domains of cytoplasmic kinases, and of phosphorylated ITIMs with the SH2 domains of cytoplasmic phosphatases, respectively.

This schematic opposition between antagonistic kinase-binding ITAMs and phosphatase-binding ITIMs was readily accepted by immunologists, and it became more and more popular as the number of ITIM-bearing molecules with negative regulatory properties increased. Knowledge accumulated during the last few years, however, has unraveled sophisticated crosstalks between ITAM- and ITIM-bearing molecules which challenge the original binary paradigm.

2 Structural and Functional Definitions of Immunoreceptor Tyrosine-based Inhibition Motifs Revisited

Since the first ITIM was identified in the intracytoplasmic domain of FcγRIIB (AMIGORENA et al. 1992; MUTA et al. 1994; DAËRON et al. 1995a), an increasing number of molecules with negative regulatory properties were described which contain one or several ITIMs in their intracytoplasmic domains (Table 1) (VIVIER and DAËRON 1997). They include all the isoforms of murine and human FcγRIIB (LATOUR et al. 1996), KIR-2D and KIR-3D (WAGTMANN et al. 1995; COLONNA and SAMARIDIS 1995), the closely related products of immunoglobulin-like transcripts (ILTs) (CELLA et al. 1997; COLONNA 1997) and leukocyte immunoglobulin-like receptors (LIRs) (COSMAN et al. 1997), the inhibitory type of paired immunoglobulin-like receptors (PIR-B) (KUBAGAWA et al. 1997), the leukocyte-associated immunoglobulin-like receptors (LAIRs) (MEYAARD et al. 1997), the inhibitory subgroup of signal-regulatory proteins (SIRP-α) (FUJIOKA et al. 1996; KHARITONENKOV et al. 1997), CD72 (WU et al. 1998; ADACHI et al. 1998), CD22 (CAMPBELL and KLINMAN 1995), gp49B1 (KATZ et al. 1996), PD-1 (NISHIMURA et al. 1998), Ly-49 (NAKAMURA et al. 1997), NKG2 A (LE DREAN et al. 1998), CD66 (HINODA et al. 1988), and MAFA (GUTHMANN et al. 1995). Most ITIM-bearing molecules are expressed by cells of hematopoietic origin with variable restrictions in their tissue distribution. FcγRIIB, ILTs/LIRs, PIR-B, and LAIRs have a wide expression; whereas KIRs, Ly-49, NKG2 A, CD22, gp49B1, CD72, PD-1, CD66, and MAFA have restricted tissue distribution. SIRP-α are remarkable in that they are expressed by hematopoietic and nonhematopoietic cells.

ITIM-bearing molecules belong to two major molecular superfamilies: the immunoglobulin superfamily (IgSF) and the C-type lectin superfamily. Most are IgSF molecules. They possess one (PD-1), two (FcγRIIB, KIR-2Ds, LAIRs, gp49B1, ILT-3, LIR-5), three (KIR-3Ds, SIRP-α), four (ILTs/LIRs), or six (PIR-B) immunoglobulin-like extracellular domains whose ligands are not all known. FcγRIIB bind complexed IgG antibodies (HULETT and HOGARTH 1994); KIRs (MORETTA et al. 1997) and LAIRs (MEYAARD et al. 1997) bind to MHC class I molecules, CD72 to CD5 (BIKAH et al. 1998), and CD22 to sialylated glycoproteins (ENGEL et al. 1995). Ly49, NKG2 A, CD66 and MAFA belong to the lectin superfamily. The extracellular domain of Ly49 binds to mouse MHC class I molecules (BROWN et al. 1997); and that of NKG2 A to HLA-E (BORREGO et al. 1998; BRAUD et al. 1998) and Qa-1 molecules (VANCE et al. 1998).

Except the FcγRIIB and the gp49B1 genes, genes encoding ITIM-bearing molecules of the IgSF belong to the a locus located on chromosome 19 in humans (19q13.1–19q13.4) (WAGTMANN et al. 1997) and on the syntenic locus on chromosome 7 in mice (Fig. 1). The FcγRIIB gene maps within the FcR locus on chromosome 1 in both humans and mice (Fig 1). (QIU et al. 1990), and gp49B1 on mouse chromosome 10 (N. Wagtmann, personal communication). Some lectin superfamily molecules are encoded by genes which map at the NKC locus on human chromosome 12 and on mouse chromosome 6 (BROWN et al. 1997). Interestingly, in genes coding for IgSF molecules, ITIMs are all encoded by a single exon that is preceded by a type 0-intron (E. Vivier and F. Vély, unpublished). This suggests that duplicated ITIM exons may have been readily translocated from gene to gene at specific loci during evolution. One may also notice that ITIM-bearing molecules such as KIRs and Ly49s, which bind specifically to highly polymorphic, classical MHC class Ia molecules, are encoded by multiallelic genes which generate molecules with extensive polymorphism of their extracellular domains (SELVAKUMAR et al. 1997). In contrast, nonclassical nonpolymorphic HLA-E/Qa-1 MHC class Ib molecules are the ligands of nonpolymorphic NKG2 A/CD94 (BORREGO et al. 1998; BRAUD et al. 1998). Likewise, FcγRIIB, which bind to the constant domains of heavy chains of different isotypes of IgG, are encoded by a single nonpolymorphic gene (QIU et al. 1990).

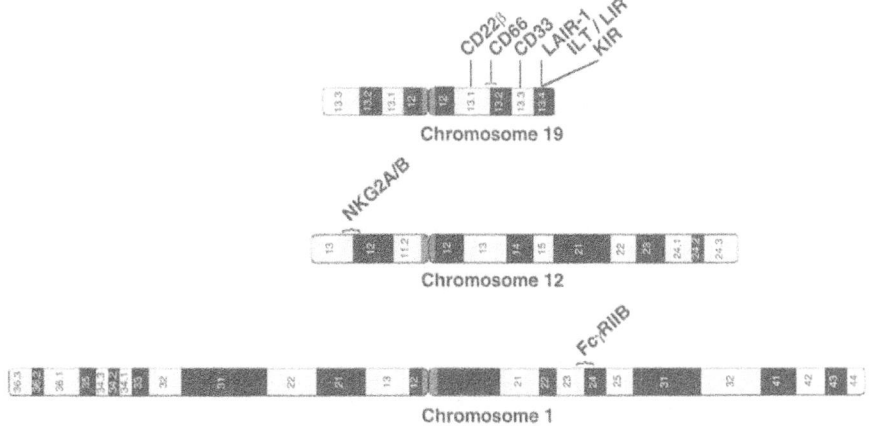

Fig. 1. Chromosome mapping of genes encoding ITIM-bearing molecules

Table 1.

Family name	Tissue distribution	Chromosome localization	Ligands	ITIM	Phosphatase(s) recruited in vivo
Immunoglobulin superfamily					
h-FcγRIIB	Basophils, monocytes, and B cells	1q23–24	IgG	AENTITYSLLMHP	?
m-FcγRIIB	Mast cells, macrophages, and B cells	1	IgG	AENTITYSLLKIHP	SHIP
h-KIR[a]	NK cells and subsets of T cells	19q13.4	HLA class I	DPQEVTYAQLNHC-x_{17}-PTDIIVYTELPNA	SHP-1/SHP-2
h-SIRPα[b]	Hematopoietic and nonhematopoietic cells	20	?	DTNDITYADLNLP-x_{11}-PNNHTEYASIQTS SEDTLTYADLDMV-x_{13}-EPSFSEYASVQVP	SHP-2/SHP-1
h-LAIR-1[c]	Monocytes and lymphocytes	19q13.4	?	SSQEVTYAQLDHW-x_{17}-MAESITYAAVARH	?
h-ILT/LIR[d]	Hematopoietic cells	19q13.4	HLA class I and?	DPQAVTYAEVKHS APQDVTYAQLHSL-x_{17}-PAEPSIYATLAIH	SHP-1
m-PIR-B	Myeloid cells	7	?	TQEESLYASVEDM-x_{16}-DPQGETYAQVKPS ESQDVTYAQLCSR-x_{17}-PEEPSVYATLAIH	SHP-1/SHP-2
m-gp49B1	Mast cells and NK cells	10	?	DPQGIVYAQVKPS-x_{10}-ETQDVTYAQLCIR	?
h-CD22β[c]	B cells	19q13.1	CD45, CD75	MDEGISYTTLRFP EDEGIHYSELIQF-x_7-AQENVDYVILKH	SHP-1
m-CD22β[c]	B cells	–	CD45, CD75	MDDTVSYAILRFP EDESIHYSELVQF-x_7-AKEDVDYVTLKH	SHP-1

hPD-1[c]	Activated lymphoid and myeloid cells	2q37	?	PVFSVDYGELDFQ-x12-VPEQTEYATIVFP	?
mPD-1[c]	Activated T and B cells	?	?	PVPSVAYEELDFQ-x10-ACVHTEYATIVFT	?
h-CD66a[c]	Neutrophils	19q13.1-2	?	KMNEVTYSTLNFE-x14-TATEIIYSEVKKQ	SHP-1
h-CD33[c]	Myeloid progenitor cells, monocytes	19q13.3	Sialoglycoconjugates	MDEELHYASLNFH-x5 -KDTSTEYSEVRTQ	?
C-type Lectin superfamily					
m-Ly49[e]	NK cells and subsets of T cells	6	H-2 class I	SEQEVTYSMVRFH	SHP-1/SHP-2
r-MAFA[c]	Mast cells	?	Oligosaccharides	MADNSIYSTLELP	ND
h-MAFA[c]	Basophils and NK cells	12p12-p13	?	MTDSVIYSMLELP	ND
h-NKG2 A/B	NK cells and subsets of T cells	12p12.3-p13.1	HLA-E	DNQGVIYSDLNLP-x19-TEQEITYAELNLQ	SHP-1/SHP-2
m-NKRP1B	NK cells and subsets of T cells	12	Carbohydrates?	DSTTLVYADLNLA	?
m-CD72	B cells	9p	CD5?	MADNSTYSTLELE-x19-EDGELTYNEVQVS	SHP-1
h-CD72	B cells	4	CD5?	MAEAITYADLRFV-x19-DDGEITYENVQVP	SHP-1

[a] KIR inhibitory molecules include at least six distinct subfamilies: KIR2DL1 (Clone 42/NKAT-1), KIR2DL2 (clone 43/NKAT-6), KIR2DL3 (clone 6/NKAT-2), KIR2DL4 (KIR103), KIR3DL1 (clone 11/NKAT-3), KIR3DL2 (clone 5/NKAT-4). Clone 6 (KIR2DL3, CD158b) is presented in this table.

[b] Human SIRPα and their activating isoforms SIRPβ include at least 15 members. SIRPα1 is presented in this table.

[c] These molecules contain putative ITIMs which have not yet been shown to act as an inhibitory motif. For CD22β, this putative ITIM (CDDTVTYSALHKR) is based on Y777.

[d] Human inhibitory ILT/LIR molecules include at least 5 distinct members (ILT2/LIR1/MIR5, ILT3/LIR5/HM18, ILT4/LIR2/MIR10, ILT5/LIR3/HL9 and LIR8). ILT2 is presented in this table.

[e] Ly49 inhibitory molecules include at least seven members. Ly49A ITIM is presented in this table.

The comparison of ITIM sequences in the many ITIM-bearing inhibitory receptors led to the recognition of a consensus motif based on three conserved residues. Classically, this motif is: YxxL/V, preceded by an aliphatic (I, V, or L) residue at position Y-2 (VIVIER and DAËRON 1997). If ITIMs are degenerate motifs, one sequence, VTYAQL, appears to be remarkably conserved among ITIM-bearing molecules of both the IgSF and the lectin superfamily with the following constant features: it is always associated with a second tandem ITIM having another sequence, from which it is separated by an average number of 30 residues; and it is located proximal to the membrane. This suggests that some ITIMs may be more canonical than others, possibly with distinct functional properties.

Their ability to recruit cytoplasmic SH2 domain-bearing phosphatases, when tyrosyl phosphorylated has been proposed as a functional definition of ITIMs (VIVIER and DAËRON 1997). The in vivo binding specificity of phosphorylated ITIMs was first defined using synthetic phosphopeptides. Most ITIMs were found to bind to SHP-1 and/or SHP-2, the two known SH2 domain-bearing cytoplasmic protein tyrosine phosphatases (CAMPBELL and KLINMAN 1995; D'AMBROSIO et al. 1995; BURSHTYN et al. 1996; OLCESE et al. 1996; KHARITONENKOV et al. 1997; VÉLY et al. 1997; ADACHI et al. 1998; WU et al. 1998). FcγRIIB and gp49B1 ITIMs bound SHP-1 and SHP-2, but also the SH2 domain-bearing inositol polyphosphate 5-phosphatase SHIP (FONG et al. 1996; ONO et al. 1996; MALBEC et al. 1998; KUROIWA et al. 1998). Studies of phosphopeptides bearing amino acid substitutions permitted identification of the aliphatic residue, at position Y-2, as determining the binding specificity of ITIMs for SHP-1 and SHP-2 (VÉLY et al. 1997; BURSHTYN et al. 1997). What determines the affinity of ITIMs for SHIP is unknown. Remarkably, ITIM-bearing molecules appeared to recruit specific phosphatases in vivo among those they can bind in vitro. Thus, in both mast cells and B cells, murine FcγRIIB selectively recruit SHIP in vivo (FONG et al. 1996; ONO et al. 1997), although, when incubated in lysates from the same cells, phosphopeptides corresponding to the FcγRIIB ITIM bind SHIP as well as SHP-1 and SHP-2 in vitro. Likewise, SHP-2 was found to be selectively recruited by the N-terminal ITIM of KIRs (BRUHNS et al. 1999), although corresponding phosphopeptides bound both SHP-1 and SHP-2 in vitro. The molecular bases for the differential selective, in vivo recruitment of phosphatases is unknown.

ITIM-bearing molecules possess variable numbers of ITIMs. FcγRIIB and Ly-49 have a single ITIM; KIRs, LAIRs, gp49B1 and Ly-49 have two ITIMS CD22 has three ITIMs; and SIRP-α, ILTs/LIRs, and PIR-B have four potential ITIMs. The biological significance of the presence of several inhibitory motifs in the majority of ITIM-bearing molecules is not fully understood. ITIMs may have redundant additive effects or nonredundant complementary and possibly cooperative effects. The N-terminal and the C-terminal ITIMs in the intracytoplasmic domain of KIRs, for example, were recently found to have nonredundant effects and to cooperate in order to fully inhibit cell activation. Thus, the N-terminal ITIM was necessary and sufficed to recruit SHP-2 in vivo, but not SHP-1, the recruitment of which required both ITIMs (BRUHNS et al. 1999).

3 A Network of Negative and Positive Receptors

One constant feature of ITIM-bearing molecules is that several such regulatory molecules are coexpressed on single cells. Thus, human NK cells express KIRs (MORETTA et al. 1997), ILT-2 (COLONNA et al. 1997), and CD94/NKG2 A (LAZETIC et al. 1996); mast cells express FcγRIIB (BENHAMOU et al. 1990), MAFA (GUT-HMANN et al. 1995), gp49B1 (CASTELLS et al. 1994), and PIR-B (KUBAGAWA et al. 1997); B cells express FcγRIIB (AMIGORENA et al. 1989), CD72 (ADACHI et al. 1998), PIR-B (KUBAGAWA et al. 1997), ILT2 (COLONNA et al. 1997), and CD22 (SATO et al. 1996). The coexpression of multiple ITIM-bearing molecules provides hematopoietic cells with multiple means of controlling cell activation at different times, in different environments, by different ligands, and by different mechanisms. Interestingly, KIRs and Ly-49 are expressed during the terminal differentiation of NK cells and of memory T cells, whereas FcγRIIB and PIR-B are expressed early on progenitors of hematopoietic cells and remain expressed on differentiated cells. Different inhibition pathways may thus be used by individual cells, at different stages of differentiation and with different biological consequences.

Positive and negative molecules which belong to different molecular families and are engaged by different ligands on single cells may indeed establish different types of crosstalk. When coaggregated with ITAM-bearing receptors, ITIM-bearing molecules which recruit SHIP (such as FcγRIIB), or SHP-1 and SHP-2 in different respective proportions (such as KIRs and SIRP-α), inhibit distinct steps of signal transduction during cell activation. Thus, FcγRIIB need to be phosphorylated by PTKs provided by activating receptors such as ITAM-bearing receptors (MALBEC et al. 1998), and they negatively regulate cell activation by interfering with downstream signaling events via SHIP (ONO et al. 1997). SHIP dephosphorylates the activation product of PI3-kinase, PtdIns(3, 4, 5)P3, which is necessary to recruit Btk to the membrane (GUPTA et al. 1997; ONO et al. 1997). Btk is mandatory for generating a sustained influx of extracellular Ca^{2+}. By contrast, KIRs may be phosphorylated by their own kinases (lck) (BINSTADT et al. 1996) and they block initial signaling events by recruiting SHP-1 and SHP-2, which may dephosphorylate ITAMs, protein tyrosine kinases, and/or LAT recruited by activating receptors (VALIANTE et al. 1996). KIRs may thus establish a potential, preexisting threshold inhibition that needs to be overcome by activating receptors, whereas FcγRIIB may ensure an "à la carte" regulation which only takes place after activation has been initiated.

Most ITIM-bearing inhibitory receptors have cell-activating counterparts which share extracellular ligands and are encoded by separate genes (VÉLY and VIVIER, 1997). Distinct genes encode FcγRIIB and FcγRIIIA (or, in humans, FcγRIIA/C) (QIU et al. 1990), inhibitory and activating KIRs (BIASSONI et al. 1996), PIR-A and PIR-B (ALLEY et al. 1998), ILT-3 and ILT-1/LIR-7 (NAKAJIMA et al. 1999), SIRP-α and SIRP-β (KHARITONENKOV et al. 1997), Ly-49 A and Ly-49D/H (TAKEI et al. 1997), and NKG2 A and NKG2 C (HOUCHINS et al. 1991). FcγRIIIA (WITHMUELLER et al. 1992), FcγRIIA/C (RETH, 1989), cell-activating

KIRs (OLCESE et al. 1997; LANIER et al. 1998b), Ly49D/H (SMITH et al. 1998), NKG2 C (LANIER et al. 1998a), and PIR-A (MAEDA et al. 1998) possess or are associated with subunits which possess ITAMs. The molecular bases of cell-activating properties of SIRP-β are not yet known. In some cases, positive and corresponding negative regulatory molecules of the same family are expressed on single cells. Thus, FcγRIIB and FcγRIIIA are coexpressed by mouse mast cells (BEN-HAMOU et al. 1990), FcγRIIB and FcγRIIA/C by human basophils (DAËRON et al. 1995a); and biological responses to the coligation of the two receptors by IgG antibodies, which cannot discriminate the two receptors, depend on the relative proportion of positive and negative receptors. The negative effect of PIR-B was found to overweigh the positive effect of PIR-A (BLÉRY et al. 1998) which are coexpressed by B cells and mast cells.

Interestingly, some ITIM-bearing molecules, such as CD22, are also activating molecules (SATO et al. 1996). This suggests that regulatory interactions involving ITIMs may take place not only "in *trans*", between distinct positive and negative molecules, but also "in *cis*", between molecular motifs with positive and negative regulatory properties respectively, and borne on the same molecule. If so, the group of ITIM-bearing molecules might be extended to include a variety of receptors, such as the erythropoietin receptor which bears ITIM-like sequences that were shown to recruit SH2 domain-bearing phosphatases which negatively regulate Jak/Stat-mediated transduction pathways (KLINGMÜLLER et al. 1995). Finally, ITIM-bearing molecules can negatively regulate not only cell activation by ITAM-bearing receptors, but also cell proliferation induced by transmembrane protein tyrosine kinase receptors (RTKs). FcγRIIB were recently found to become tyrosyl phosphorylated, to recruit SHIP, and to inhibit c-kit-mediated proliferation of mouse mast cells when coligated with c-kit by IgG anti-c-kit antibodies. Inhibition was correlated with a blockade of the cell cycle in the G1 phase (MALBEC et al. 1999). Likewise, the overexpression of SIRP-α was sufficient to inhibit RTK-dependent cell proliferation via the recruitment of SHP-2 (KHARITONENKOV et al. 1997). These observations further extend the list of biological processes that can be controlled by ITIM-bearing receptors.

4 Concluding Remarks

A complex network involving ITIM-bearing molecules and a variety of receptors capable of triggering cell activation and/or cell proliferation and differentiation may thus take place at the surface of a large number of cells. These include, but are not restricted to, cells of hematopoietic origin. Depending on the cell type, their differentiation stage, and available extracellular ligands, different configurations of this network can be established. Such networks may be more intricate, since molecules identified as being involved in regulation by negative receptors appear to regulate constitutively cell-activating receptors. SHIP, for example, can be recruited not only by FcγRIIB, but also by FcεRI, and may negatively regulate IgE-dependent mast cell activation in the absence of IgG. Indeed, mast cells from SHIP-deficient mice exhibited enhanced IgE-induced mediator release (HUBER et al. 1998). It follows that the opposition between negative and positive receptors may turn out to be more apparent than real and that the latter molecules could exert more subtle regulatory effects than was anticipated from the original all-or-none paradigm.

Our aim was to invite colleagues who have made significant contributions in the field of ITIMs to write an updated review on specific ITIM-bearing molecules and to have all the reviews published together for the first time in a single volume. Some ITIM-bearing molecules, such as SIRP-α, CD66, PD1 or CD72, are not reviewed in this volume. We nevertheless believe that the present set of papers on the best known ITIMs, viewed from different perspectives, might provide the basis for an integrated scheme that we hope will contribute to a better understanding of the biology of ITIM-bearing molecules.

Acknowledgments. We are grateful to Wolf H. Fridman for his critical reading, and to Frédéric Vély and Mathieu Bléry for their help in the completion of this manuscript.

References

Adachi T, Flaswinkel H, Yakura H, Reth M, Tsubata T (1998) The B cell surface protein CD72 recruits the tyrosine phosphatase SHP-1 upon tyrosine phosphorylation. J Immunol 160:4662–4665

Alley TL, Cooper MD, Chen M, Kubagawa H (1998) Genomic structure of PIR-B, the inhibitory member of the paired immunoglobulin-like receptor genes in mice. Tissue Antigens 51:224–231

Amigorena S, Bonnerot C, Choquet D, Fridman WH, Teillaud JL (1989) FcγRII expression in resting and activated B lymphocytes. Eur J Immunol 19:1379–1385

Amigorena S, Bonnerot C, Drake J, Choquet D, Hunziker W, Guillet JG, Webster P, Sautès C, Mellman I, Fridman WH (1992) Cytoplasmic domain heterogeneity and functions of IgG Fc receptors in B-lymphocytes. Science 256:1808–1812

Benhamou M, Bonnerot C, Fridman WH, Daëron M (1990) Molecular heterogeneity of murine mast cell Fcγ receptors. J Immunol 144:3071–3077

Biassoni R, Cantoni C, Falco M, Verdiani S, Bottino C, Vitale M, Conte R, Poggi A, Moretta A, Moretta L (1996) The human leukocyte antigen (HLA)-C-specific "activatory" or "inhibitory" natural killer cell receptors display highly homologous extracellular domains, but differ in their transmembrane and intracytoplasmic portions. J Exp Med 183:645–650

Bikah G, Lynd FM, Aruffo AA, Ledbetter JA, Bondada S (1998) A role for CD5 in cognate interactions between T cells and B cells, and identification of a novel ligand for CD5. Int Immunol 10:1185–1196

Binstadt BA, Brumbaugh KM, Dick CJ, Scharenberg AM, Williams BL, Colonna M, Lanier LL, Kinet J-P, Abraham RT, Leibson PJ (1996) Sequential involvement of lck and SHP-1 with MHC-recognizing receptors on NK cells inhibits FcR-initiated tyrosine kinase activation. Immunity 5:629–638

Bléry M, Delon J, Trautmann A, Cambiaggi A, Olcese L, Biassoni R, Moretta L, Chavrier P, Moretta A, Daëron M, Vivier E (1997) Reconstituted killer-cell inhibitory receptors for MHC class I molecules control mast cell activation induced via immunoreceptor tyrosine-based activation motifs. J Biol Chem 272:8989–8996

Bléry M, Kubagawa H, Chen CC, Vely F, Cooper MD, Vivier E (1998) The paired Ig-like receptor PIR-B is an inhibitory receptor that recruits the protein-tyrosine phosphatase SHP-1. Proc Natl Acad Sci USA 95:2446–2451

Borrego F, Ulbrecht M, Weiss EH, Coligan JE, Brooks AG (1998) Recognition of human histocompatibility leukocyte antigen (HLA)-E complexed with HLA class I signal sequence-derived peptides by CD94/NKG2 confers protection from natural killer cell-mediated cytotoxicity. J Exp Med 187:813–818

Braud VM, Allan DSJ, O'Callaghan CA, Söderström K, D'Andrea A, Ogg GS, Lazetic S, Young NT, Bell JI, Phillips JH, Lanier LL, McMichael AJ (1998) HLA-E binds to natural killer cell receptors CD94/NKG2 A, B and C. Nature 391:795–799

Brown MG, Scalzo AA, Matsumoto K, Yokoyama WM (1997) The natural killer cell gene complex: a genetic basis for understanding natural killer cell function and innate immunity. Immunol Rev 155:53–65

Bruhns P, Marchetti P, Fridman WH, Vivier E, Daëron M (1999) Differential roles of N- and C-terminal ITIMs during inhibition of cell activation by killer cell inhibitory receptors. J Immunol 162: 3168–3175

Burshtyn DN, Long EO (1997) Regulation through inhibitory receptors: lessons from killer cells. Trends Cell Biol 7:473–479

Burshtyn DN, Scharenberg AM, Wagtmann N, Rajogopalan S, Berrada K, Yi T, Kinet J-P, Long EO (1996) Recruitment of tyrosine phosphatase HCP by the killer cell inhibitory receptor. Immunity 4:77–85

Burshtyn DN, Yang W, Yi T, Long EO (1997) A novel phosphotyrosine motif with a critical amino acid at position –2 for the SH2 domain-mediated activation of the tyrosine phosphatase SHP-1. J Biol Chem 272:13066–13072

Cambier JC (1995) Antigen and Fc Receptor signaling: The awesome power of the immunoreceptor tyrosine-based activation motif (ITAM). J Immunol 155:3281–3285

Campbell M-A, Klinman NR (1995) Phosphotyrosine-dependent association between CD22 and protein tyrosine phosphatase 1 C. Eur J Immunol 25:1573–1579

Castells MC, Wu X, Arm JP, Austen KF, Katz HR (1994) Cloning of the gp49B gene of the immunoglobulin superfamily and demonstration that one of its two products is an early-expressed mast cell surface protein originally described as gp49. J Biol Chem 269:8393–8401

Cella M, Döhring C, Samaridis J, Dessing M, Brockhaus M, Lanzavecchia A, Colonna M (1997) A novel inhibitory receptor (ILT3) expressed on monocytes, macrophages, and dendritic cells involved in antigen processing. J Exp Med 185:1743–1751

Colonna M (1997) Immunoglobulin-superfamily inhibitory receptors: from natural killer cells to antigen-presenting cells. Research Immunol 148:168–171

Colonna M, Navarro F, Bellon T, Llano M, Garcia P, Samaridis J, Angman L, Cella M, Lopez-Botet M (1997) A common inhibitory receptor for major histocompatibility complex class I molecules on human lymphoid and myelomonocytic cells. J Exp Med 186:1809–1818

Colonna M, Samaridis J (1995) Cloning of immunoglobulin-superfamily members associated with HLA-C and HLA-B recognition by human natural killer cells. Science 268:405–408

Cosman D, Fanger N, Borges L, Kubin M, Chin W, Peterson L, Hsu ML (1997) A novel immunoglobulin superfamily receptor for cellular and viral MHC class I molecules. Immunity 7:273–282

D'Ambrosio D, Hippen KH, Minskoff SA, Mellman I, Pani G, Siminovitch KA, Cambier JC (1995) Recruitment and activation of PTP1 C in negative regulation of antigen receptor signaling by FcγRIIB1. Science 268:293–296

Daëron M, Latour S, Malbec O, Espinosa E, Pina P, Pasmans S, Fridman WH (1995a) The same tyrosine-based inhibition motif, in the intracytoplasmic domain of FcγRIIB, regulates negatively BCR-, TCR-, and FcR-dependent cell activation. Immunity 3:635–646

Daëron M, Malbec O, Latour S, Aroc M, Fridman WH (1995b) Regulation of high-affinity IgE receptor-mediated mast cell activation by murine low-affinity IgG receptors. J Clin Invest 95:577–585

Engel P, Wagner N, Miller A, Tedder TF (1995) Identification of the ligand-binding domains of CD22, a member of the immunoglobulin superfamily that uniquely binds a sialic acid-dependent ligand. J Exp Med 181:1581–1586

Fong DC, Malbec O, Arock M, Cambier JC, Fridman WH, Daëron M (1996) Selective in vivo recruitment of the phosphatidylinositol phosphatase SHIP by phosphorylated FcγRIIB during negative regulation of IgE-dependent mouse mast cell activation. Immunol Letters 54:83–91

Fujioka Y, Matozaki T, Noguchi T, Iwamatsu A, Yamao T, Takahashi N, Tsuda M, Takada T, Kasuga M (1996) A novel membrane glycoprotein, SHPS-1, that binds the SH2-domain-containing protein tyrosine phosphatase SHP-2 in response to mitogens and cell adhesion. Mol Cell Biol 16:6887–6899

Gupta N, Scharenberg AM, Burshtyn DN, Wagtmann N, Lioubin MN, Rohrschneider LR, Kinet JP, Long EO (1997) Negative signaling pathways of the killer cell inhibitory receptor and Fc gamma RIIb1 require distinct phosphatases. J Exp Med 186:473–478

Guthmann MD, Tal M, Pecht I (1995) A secretion inhibitory signal transduction molecule on mast cells is another C-type lectin. Proc Natl Acad Sci USA 92:9397–9401

Hinoda Y, Neumaier M, Hefta SA, Drzeniek Z, Wagener C, Shively L, Hefta LJ, Shively JE, Paxton RJ (1988) Molecular cloning of a cDNA coding biliary glycoprotein I: primary structure of a glycoprotein immunologically crossreactive with carcinoembryonic antigen. Proc Natl Acad Sci USA 85:6959–6963

Houchins JP, Yabe T, McSherry C, Bach FH (1991) DNA sequence analysis of NKG2, a family of related cDNA clones encoding type II integral membrane proteins on human natural killer cells. J Exp Med 173:1017–1020

Huber M, Helgason CD, Damen JE, Liu L, Humphries RK, Krystal G (1998) The src homology 2-containing inositol phosphatase (SHIP) is the gatekeeper of mast cell degranulation. Proc Natl Acad Sci USA 95:11330–11335

Hulett MD, Hogarth PM (1994) Molecular basis of Fc Receptor function. Adv Immunol 57:1–127

Katz HR, Vivier E, Castells MC, McCormick MJ, Chambers JM, Austen KF (1996) Mouse mast cell gp49B1 contains two immunoreceptor tyrosine-based inhibition motifs and suppresses mast cell activation upon coligation with FcεRI. Proc Natl Acad Sci USA 93:10809–10814

Kharitonenkov A, Chen Z, Sures I, Wang H, Schilling J, Ullrich A (1997) A family of proteins that inhibit signalling through tyrosine kinase receptors. Nature 386:181–186

Klingmüller U, Lorenz U, Cantley LC, Neel BG, Lodish HF (1995) Specific recruitment of SH-PTP1 to the erythropoietin receptor causes inactivation of Jak2 and termination of proliferative signals. Cell 80:729–738

Kubagawa H, Burrows PD, Cooper MD (1997) A novel pair of immunoglobulin-like receptors expressed by B cells and myeloid cells. Proc Natl Acad Sci USA 94:5261–5266

Kuroiwa A, Yamashita Y, Inui M, Yuasa T, Ono M, Nagabukuro A, Matsuda Y, Takai T (1998) Association of tyrosine phosphatases SHP-1 and SHP-2, inositol 5-phosphatase SHIP with gp49B1, and chromosomal assignment of the gene. J Biol Chem 273:1070–1074

Lanier LL, Corliss B, Wu J, Phillips JH (1998a) Association of DAP-12 with activating CD94/NKG2 C NK cell receptors. Immunity 8:693–701

Lanier LL, Corliss BC, Wu J, Leong C, Phillips JH (1998b) Immunoreceptor DAP12 bearing a tyrosine-based activation motif is involved in activating NK cells. Nature 391:703–707

Latour S, Fridman WH, Daëron M (1996) Identification, molecular cloning, biological properties and tissue distribution of a novel isoform of murine low-affinity IgG receptor homologous to human FcγRIIB1. J Immunol 157:189–197

Law C-L, Sidorenko SP, Chandra KA, Zhao Z, Shen S-H, Fischer EH, Clark AA (1996) CD22 associates with protein tyrosine phosphatase 1 C, syk and phospholipase C-γ1 upon B cell activation. J Exp Med 183:547–560

Lazetic S, Chang C, Houchins JP, Lanier LL, Phillips JH (1996) Human natural killer cell receptors involved in MHC class I recognition are disulfide-linked heterodimers of CD94 and NKG2 subunits. J Immunol 157:4741–4745

Le Drean E, Vély F, Olcese L, Cambiaggi A, Guia S, Krysta G, Gervois N, Moretta A, Jotereau F, Vivier E (1998) Inhibition of antigen-induced T cell response and antibody-induced NK cell cytotoxicity by NKG2 A: association of NKG2 A with SHP-1 and SHP-2 protein-tyrosine phosphatases. Eur J Immunol 28:264–276

Maeda A, Kurosaki M, Kurosaki T (1998) Paired immunoglobulin-like receptor (PIR)-A is involved in activating mast cells through its association with Fc receptor gamma chain. J Exp Med 188:991–995

Malbec O, Fong D, Turner M, Tybulewicz VLJ, Cambier JC, Fridman WH, Daëron M (1998) FcεRI-associated lyn-dependent phosphorylation of FcγRIIB during negative regulation of mast cell activation. J Immunol 160:1647–1658

Malbec O, Fridman WH, Daëron M (1999) Negative regulation of c-kit-mediated cell proliferation by FcγRIIB. J Immunol 162:4424–4429

Meyaard L, Adema G, Chang C, Woollatt E, Sutherland GR, Lanier LL, Phillips JH (1997) LAIR-1, a novel inhibitory receptor expressed on human mononuclear leukocytes. Immunity 7:283–290

Moretta A, Biassoni R, Bottino C, Pende D, Vitale M, Poggi A, Mingari MC, Moretta L (1997) Major histocompatibility complex class I-specific receptors on human natural killer and T lymphocytes. Immunol Rev 155:105–117

Muta T, Kurosaki T, Misulovin Z, Sanchez M, Nussenzweig MC, Ravetch JV (1994) A 13-amino-acid motif in the cytoplasmic domain of FcγRIIB modulates B-cell receptor signalling. Nature 368:70–73

Nakajima H, Samaridis J, Angman L, Colonna U (1999) Human Myeloid cells express an activating ILT receptor (ILT-1) that associates with Fc receptor gamma chain. J Immunol 162:5–8

Nishimura H, Nagahiro M, Nakano T, Honjo T (1998) Immunological studies on PD-1-deficient mice: implication of PD-1 as a negative regulator for B cell responses. Int Immunol 10:1563–1572

Olcese L, Cambiaggi A, Semenzato G, Bottino C, Moretta A, Vivier E (1997) Human killer cell activatory receptors for MHC class I molecules are included in a multimeric complex expressed by natural killer cells. J Immunol 158:5083–5086

Olcese L, Lang P, Vély F, Cambiaggi A, Marguet D, Bléry M, Hippen KL, Biassoni R, Moretta A, Moretta L, Cambier JC, Vivier E (1996) Human and mouse killer-cell inhibitory receptors recruit PTP1 C and PTP1D protein tyrosine phosphatases. J Immunol 156:4531–4534

Ono M, Bolland S, Tempst P, Ravetch JV (1996) Role of the inositol phosphatase SHIP in negative regulation of the immune system by the receptor FcγRIIB. Nature 383:263–266

Ono M, Okada H, Bolland S, Yanagi S, Kurosaki T, Ravetch JV (1997) Deletion of SHIP or SHP-1 reveals two distinct pathways for inhibitory signaling. Cell 90:293–301

Pessino A, Sivori S, Bottino C, Malaspina A, Morelli L, Moretta L, Biassoni R, Moretta A (1998) Molecular Cloning of NKp46: A novel member of the immunoglobulin superfamily involved in triggering of natural cytotoxicity. J Exp Med 188:953–960

Qiu WQ, de Bruin D, Brownstein BH, Pearse R, Ravetch JV (1990) Organization of the human and mouse low-affinity FcγR genes : duplication and recombination. Science 248:732–735

Renard V, Cambiaggi A, Vely F, Blery M, Olcese L, Olivero S, Bouchet M, Vivier E (1997) Transduction of cytotoxic signals in natural killer cells: a general model of fine tuning between activatory and inhibitory pathways in lymphocytes. [Review] [153 refs]. Immunol Rev 155:205–221

Reth MG (1989) Antigen receptor tail clue. Nature 338:383–384

Sato S, Miller AS, Inaoki M, Bock CB, Jansen PJ, Tang MLK, Tedder TF (1996) CD22 is both a positive and negative regulator of B lymphocyte antigen receptor signal transduction: Altered signaling in CD22-deficient mice. Immunity 5:551–562

Selvakumar A, Steffens U, Dupont B (1997) Polymorphism and domain variability of human killer cell inhibitory receptors. Immunol Rev 155:183–196

Smith KM, Wu J, Bakker ABH, Phillips JH, Lanier LL (1998) Ly-49D and Ly-49H associate with mouse DAP12 and form activating receptors. J Immunol 161:7–10

Takei F, Brennab J, Mager, DL (1997) The Ly-49 family: genes, proteins, and recognition of class I MHC. Immunol Rev 155:67–78

Valiante NM, Phillips JH, Lanier LL, Parham P (1996) Killer cell inhibitory receptor recognition of human leukocyte antigen (HLA) class I blocks formation of a pp36/PLC- signaling complex in human natural killer (NK) Cells. J Exp Med 184:2243–2250

Vance RE, Kraft JR, Altman JD, Jensen PE, Raulet DH (1998) Mouse CD94/NKG2 A is a natural killer cell receptor for the nonclassical major histocompatibility complex (MHC) class I molecule Qa-1. J Exp Med 188:1841–1848

Vély F, Olivero S, Olcese L, Moretta A, Damen JE, Liu L, Krystal G, Cambier JC, Daëron M, Vivier E (1997) Differential association of phosphatases with hematopoietic coreceptors bearing immunoreceptor tyrosine-based inhibition motifs. Eur J Immunol 27:1994–2000

Vély F, Vivier E (1997) Conservation of structural features reveals the existence of a large family of inhibitory cell surface receptors and noninhibitory/activatory counterparts. J Immunol 159: 2075–2077

Vivier E, Daëron M (1997) Immunoreceptor tyrosine-based inhibition motifs. Immunol Today 18: 286–291

Wagtmann N, Biassoni R, Cantoni C, Verdiani S, Malnati M, Vitale M, Bottino C, Moretta L, Moretta A, Long EO (1995) Molecular clones of p58 NK cell receptor reveal immunoglobulin-related molecules with diversity in both the extra- and the intracellular domains. Immunity 2:439–449

Wagtmann N, Rojo S, Eichler E, Mohrenweiser H, Long EO (1997) A new human gene complex encoding the killer cell inhibitory receptors and related monocyte/macrophage receptors. Curr Biol 8:1009–1017

Wang LL, Mehta IK, LeBlanc PA, Yokoyama WM (1997) Mouse natural killer cells express gp49B1, a structural homologue of human killer cell inhibitory receptors. J Immunol 158:13–17

Withmueller U, Kurosaki T, Murakami MS, Ravetch JV (1992) Signal transduction by FcγRIII (CD16) is mediated through the γ chain. J Exp Med 175:1381–1390

Wu Y, Nadler MJS, Brennan LA, Gish GD, Timms JF, Fusaki N, Jongstra-Bilen J, Tada N, Pawson T, Wither J, Neel BG, Hozumi N (1998) The B-cell transmembrane protein CD72 binds to and is an in vivo substrate of the protein tyrosine phosphatase SHP-1. Curr Biol 8:1009–1017

Negative Regulation of Hematopoietic Cell Activation and Proliferation by FcγRIIB

O. Malbec, W.H. Fridman, and M. Daëron

1 Introduction

FcγRIIB are single-chain, low-affinity receptors for the Fc portion of IgG which belong to the immunoglobulin superfamily (Daëron 1997a). They bind IgG antibodies either complexed to multivalent soluble antigens or bound to cell membrane antigens. The FcγRIIB gene possesses three exons that encode the intracytoplasmic domain. By alternative splicing, it generates two (in humans) or three (in mice) isoforms of membrane FcγRIIB that differ only in their intracytoplasmic sequences (Ravetch et al. 1986; Hibbs et al. 1986; Lewis et al. 1986; Brooks et al. 1989; Latour et al. 1996). FcγRIIB are involved in the internalization of IgG immune complexes; and FcγRIIB isoforms were shown to differentially mediate the phagocytosis of particulate immune complexes and/or the endocytosis of soluble immune complexes (Daëron et al. 1993). No FcγRIIB isoform is able to trigger cell activation (Daëron et al. 1992). Instead, all isoforms were found to negatively regulate cell activation.

Laboratoire d'Immunologie Cellulaire et Clinique, INSERM U.255, Institut Curie, Paris, France

2 FcγRIIB as General Negative Regulators of Hematopoietic Cells

Evidence that receptors for the Fc portion of IgG (FcγR) can negatively regulate cell activation stems from early in vivo works which showed that passively administered IgG antibodies to an immunogen specifically suppressed antibody responses to that immunogen (HENRY and JERNE 1968). Suppression was also induced by IgG antibodies to epitopes borne by immunoglobulins (MANNING and JUTILA 1972; SHEK and DUBISKI 1975; KOHLER et al. 1978). In vivo and in vitro analysis showed that the suppressive properties of antibodies require an intact Fc portion (SINCLAIR and CHAN 1971; SHEK and DUBISKI 1975; KOHLER et al. 1977; HEYMAN 1990), and identified surface immunoglobulins on B cells, which were later shown to constitute the antigen-binding subunit of the B cell receptor (BCR), as the targets of suppressive antibodies or antigen–antibody complexes. Based on these findings, a model was proposed by Sinclair, according to which the coaggregation of surface immunoglobulins and FcγR expressed by B cells delivered a negative signal that could inhibit the positive signal delivered by the aggregation of BCR (SINCLAIR and CHAN 1971). Involvement of FcγR in negative signaling was subsequently confirmed by showing that monoclonal antibodies to the binding site of mouse FcγR prevented anti-immunoglobulin-induced inhibition of B cell activation in vitro (PHILLIPS and PARKER 1984) and in vivo (HEYMAN 1989). Later on, when FcγR were cloned, mouse B cells were shown to express no FcγR other than FcγRIIB (AMIGORENA et al. 1989; LATOUR et al. 1996). Reconstitution experiments conducted in the IIA1.6 cell line, a FcγRIIB-deficient variant of the murine B cell lymphoma A20/2 J, showed that the inhibitory properties of FcγRIIB reside in sequences of the intracytoplasmic domain encoded by the third intracytoplasmic exon that are common to all three murine FcγRIIB isoforms (AMIGORENA et al. 1992; LATOUR et al. 1996). It then became a popular paradigm that B cell activation could be negatively regulated, through FcγRIIB, by the end-products of B cell activation (i.e. IgG antibodies) which, in the form of immune complexes, could coaggregate BCR via the antigen moiety and FcγRIIB via the Fc portion of antibodies.

FcγRIIB, however, are expressed by hematopoietic cells other than B lymphocytes. Murine FcγRIIB are also expressed by activated T cells, particularly of the TH2 and γ/δ subsets (SANDOR and LYNCH 1992), macrophages, monocytes (MELLMAN et al. 1988), Langerhans cells (ESPOSITO-FARESE et al. 1995), and mast cells (BENHAMOU et al. 1990). Human FcγRIIB are constitutively expressed by macrophages, monocytes, and Langerhans cells (HULETT and HOGARTH 1994). They are also expressed on some activated T cells (MANTZIORIS et al. 1993). FcγRIIB are expressed early in ontogeny (LYNCH and SANDOR 1990) on precursor cells of the lymphoid, myeloid, and megakaryocytic lineages. FcγRIIB are expressed on fetal hematopoietic cells, at a time when antibody synthesis has not yet started, but when maternal antibodies are still present. As a consequence, IgG antibodies could potentially negatively regulate the activation of FcγRIIB-positive hematopoietic cells other than B cells.

The recognition of a shared immunoreceptor tyrosine-based activation motif (ITAM) in the intracytoplasmic domains of transduction subunits associated with the BCR, the T cell receptor (TCR), and FcR (RETH 1989) provided the molecular basis for an extension of the negative regulatory properties of FcγRIIB to cell activation by receptors other than BCR. Indeed, ITAM-bearing receptors were shown to trigger cell activation by using variations on a common signal transduction scheme (CAMBIER 1995). Immunoreceptors have no intrinsic enzymatic activity, but they are constitutively or inducibly associated with one or several membrane-anchored protein tyrosine kinases of the src family. Upon receptor aggregation, these kinases tyrosyl phosphorylate ITAMs. Phosphorylated ITAMs then provide docking sites for SH2 domain-bearing cytoplasmic molecules which initiate chains of enzymatic reactions. Recruited molecules are primarily cytoplasmic protein tyrosine kinases (Syk or ZAP70) that lead to the mobilization of intracellular Ca^{2+}, and adapter molecules, such as shc, that connect immunoreceptors to metabolic pathways, such as the ras pathway, leading to the expression of inducible genes (DAËRON 1997a).

Based on these grounds, we investigated whether FcγRIIB could affect cell activation by ITAM-bearing receptors other than BCR. We found that, besides BCR-mediated B cell activation, FcγRIIB could also negatively regulate TCR-mediated T cell activation and FcR-mediated mast cell activation. Using single-chain chimeric molecules, the intracytoplasmic domain of which contained a single ITAM derived from the BCR-associated Igα subunit (MUTA et al. 1994), the TCR-associated ζ subunit, or the FcεRI-associated FcRγ subunit (DAËRON et al. 1995a), and which were coaggregated with FcγRIIB expressed on B cells or mast cells, FcγRIIB-mediated negative regulation was extended to all ITAM-bearing receptors (DAËRON et al. 1995a). Remarkably, in order to inhibit cell activation, FcγRIIB needed to be coaggregated with ITAM-bearing receptors by the same bispecific extracellular ligand, whatever its fine specificity, and cell activation needed to be triggered via those receptors that were coaggregated with FcγRIIB (DAËRON et al. 1995b).

Whether FcγRIIB-mediated negative regulation was specific for ITAM-dependent cell activation or whether it could be further extended to cell activation by receptors with no ITAM remained an unanswered question. From a teleological point of view, it might seem logical that one of the major effector molecules of the immune system, i.e., IgG antibodies, would specifically control the activation of cells bearing receptors involved in the recognition of antigen under its different forms. IgG autoantibodies to a variety of cell surface molecules were, however, found in the serum of normal individuals (AVRAMEAS et al. 1981) and in the serum of patients suffering from various diseases (KAZATCHKINE et al. 1994). We therefore examined whether FcγRIIB might inhibit mast cell proliferation induced via c-kit, a model of growth factor receptor tyrosine kinase (RTK) of the PDGF receptor family which controls mast cell development and differentiation (DUBREUIL et al. 1990). By contrast with immunoreceptors, RTKs have an intrinsic enzymatic activity due to the presence of a protein tyrosine kinase domain in their intracytoplasmic portion. When dimerized by growth factors – stem cell factor (SCF) in the case of c-kit – they autophosphorylate and recruit cytoplasmic signaling molecules which lead to the entry of cells into the cell cycle and, ultimately, to cell division. Among the molecules recruited by RTKs, several are also recruited by ITAM-bearing receptors (ROTTAPEL et al. 1991).

We found that, when complexed to F(ab')2 fragments of polyclonal mouse anti-rat immunoglobulins, F(ab')2 fragments of a rat anti-mouse c-kit (OGAWA et al. 1991) could mimic the effects of SCF and induce [^{3}H]thymidine incorporation in growth-factor-dependent bone-marrow-derived mouse mast cells which constitutively express FcγRIIB. Under the same conditions, intact IgG anti-c-kit failed to induce significant thymidine incorporation, unless mast cells were preincubated with 2.4G2, a rat monoclonal antibody which blocks the binding site of mouse FcγRIIB, or were derived from FcγRIIB-deficient mice. C-kit-mediated mast cell proliferation can therefore be negatively regulated by FcγRIIB, and inhibition was correlated with an arrest of the cell cycle in G1 (MALBEC et al. 1999). As other RTKs utilize similar transduction pathways as c-kit to deliver proliferation signals to cells, one can hypothesize that FcγRIIB may control cell growth triggered by other RTKs. These include platelet-derived growth factor (PDGF) receptors, colony-stimulating factor (CSF) receptors, epithelial growth factor (EGF) receptors, fibroblast growth factor (FGF) receptors, nerve growth factor (NGF) receptors, vascular endothelial growth factor (VEGF) receptors, insulin-like growth factor (IGF) receptors, and insulin receptors. Major effector molecules of the immune system, IgG antibodies, may thus control the proliferation of most cells of hematopoietic lineage.

FcγRIIB can therefore be understood now as general regulators not only of ITAM-bearing receptor-dependent activation of immune cells, but also of growth factor receptor-dependent proliferation of hematopoietic cells.

3 FcγRIIB and the Concept of Immunoreceptor Tyrosine-based Inhibition Motif

The mutational analysis of sequences responsible for FcγRIIB-mediated inhibition of BCR-, TCR-, and FcR-dependent cell activation identified a 13-amino acid sequence that was necessary (AMIGORENA et al. 1992; DAËRON et al. 1995a) and sufficient (MUTA et al. 1994) for inhibition. The point mutation of the one tyrosine residue contained in that sequence sufficed to abrogate the inhibitory properties of FcγRIIB (MUTA et al. 1994; DAËRON et al. 1995a). Because this tyrosine is followed, at position Y + 3, by a leucine residue, the YxxL motif was reminiscent of the double YxxL motif of ITAMs, and the term "immunoreceptor tyrosine-based inhibition motif" (ITIM) was adopted at a time when it was known to be present in a single molecule only. The subsequent recognition of similar motifs first in killer cell inhibitory receptors (KIRs) (OLCESE et al. 1996; BURSHTYN et al. 1997), then in an increasing number of molecules with negative regulatory properties (UNKELESS and JIN 1997; DAËRON 1997b), substantiated the concept of an inhibitory motif; and sequence comparison of ITIM-bearing molecules revealed the conservation of a third residue (I, L or V) at position Y-2 (BURSHTYN et al. 1997; VIVIER and DAËRON 1997).

Like ITAMs, ITIMs are not functional as such: they need to become tyrosyl phosphorylated and, like ITAMs, when phosphorylated, they recruit cytoplasmic molecules bearing SH2 domains. The ability of phosphorylated ITIMs to bind SH2 domain-bearing molecules in vitro was first analyzed using immobilized synthetic peptides incubated with lysates from [^{35}S]methionine-labeled cells by autoradiography and Western blot analysis (D'AMBROSIO et al. 1995; FONG et al. 1996; BURSHTYN et al. 1997; MALBEC et al. 1998). Remarkably, phospho-ITIMs were found to have an affinity for SH2 domain-bearing phosphatases only. All phospho-ITIMs studied so far had an affinity for one or both protein tyrosine phosphatases, SHP-1 and SHP-2 (SCHARENBERG and KINET 1996; UNKELESS and JIN 1997; LONG et al. 1997). A few, including the FcγRIIB ITIM, also bound the phosphatidylinositol polyphosphate 5-phosphatase SHIP (ONO et al. 1996). The Y-2 residue was shown to determine the ability of phospho-ITIMs to bind SHP-1 and SHP-2, and to have no influence on the binding of SHIP (VÉLY et al. 1997).

The original intuition that the inhibitory sequence of FcγRIIB might contain an inhibitory motif with a wide distribution thus proved to be amply validated when other molecules with negative regulatory properties were identified. Ironically, however, the first recognized ITIM turned out to be a unique inhibitory motif, and FcγRIIB were found to be unique inhibitory receptors among all ITIM-bearing molecules. Indeed, the mechanism by which FcγRIIB negatively regulate cell activation and proliferation is not the same as mechanisms used by other ITIM-bearing negative receptors.

4 Mechanism of FcγRIIB-dependent Negative Regulation

The mechanism of FcγRIIB-mediated inhibition has been extensively studied in B lymphocytes. B lymphocytes, however, present various disadvantages. The activation of B lymphocytes induced by BCR aggregation involves several coreceptors, such as CD22 and CD19, that are constitutively associated with BCR (PLEINMAN et al. 1994). Most studies of FcγRIIB-mediated inhibition in B lymphocytes compared the consequences of aggregating BCR, using F(ab')2 fragments of anti-BCR antibodies, with the effects of coaggregating FcγRIIB with BCR, using intact IgG anti-BCR. Mast cells provide a simpler model than B cells. FcεRI are not known to be associated with coreceptors in mast cells, and reconstitution experiments in fibroblasts suggested that they need no coreceptor to transduce activation signals (SCHARENBERG et al. 1995). FcεRI aggregation leads to the release of preformed granule mediators within minutes, followed by the release of newly formed lipid-derived mediators and, ultimately, by the secretion of cytokines. Release of pre-formed mediators is a direct consequence of increased intracellular Ca^{2+} concentration, whereas cytokine synthesis also requires the activation of the ras pathway. To study the mechanism of FcγRIIB-mediated mast cell inhibition, we designed a combination of ligands which enabled us to examine intracellular events following the aggregation of FcγRIIB, the aggregation of FcεRI, or the coaggregation of FcγRIIB with FcεRI. Using these ligands, we demonstrated that the simple occupancy of FcγRIIB or the simultaneous aggregation of FcγRIIB do not prevent IgE-induced mediator release by mast cells, whereas the coaggregation of FcεRI with FcγRIIB inhibits the release of granule mediators and the secretion of cytokines (DAËRON et al. 1995b). Inhibition was correlated with a decreased extracellular Ca^{2+} influx (FONG et al. 1996; MALBEC et al. 1998). Likewise, the inhibition of c-kit-dependent mast cell proliferation required the coaggregation of c-kit with FcγRIIB (MALBEC et al. 1999).

We found that the coaggregation of FcεRI with FcγRIIB did not affect the phosphorylation of FcεRI ITAMs. As this phosphorylation was shown to be mediated by the PTK lyn (SCHARENBERG et al. 1995), one can assume that lyn is normally recruited and/or activated upon FcγRIIB-FcεRI coaggregation. We also found that syk was normally recruited and activated upon FcγRIIB-FcεRI coaggregation (MALBEC et al. 1998), indicating that FcγRIIB does not affect the initials steps of signal transduction by FcεRI. When coaggregated with FcεRI, FcγRIIB was phosphorylated on the ITIM tyrosine within seconds. Since FcγRIIB are not phosphorylated when aggregated, they are probably not constitutively associated with a PTK. We therefore hypothesized that FcγRIIB might be phosphorylated by FcεRI-associated PTKs. By in vitro kinase assays, we showed that PTKs which coprecipitate with FcεRI and are activated upon FcεRI aggregation, could indeed phosphorylate a GST fusion protein made of the intracytoplasmic domain of FcγRIIB. When immunoprecipitated from activated mast cells, either lyn or syk could phosphorylate in vitro the same fusion protein. However, when coaggregated with FcεRI, FcγRIIB became tyrosyl phosphorylated in mast cells derived from syk-deficient mice, as efficiently as in wild-type mast cells, but not in mast cells derived from lyn-deficient mice (MALBEC et al. 1998). In mast cells, the in vivo phosphorylation of FcγRIIB, therefore, does not depend on syk, but requires lyn; and lyn cannot be replaced by another PTK. In contrast, FcγRIIB were normally phosphorylated upon coaggregation with BCR in lyn-deficient B lymphocytes, suggesting that lyn can be replaced by another PTK in B cells (WANG et al. 1996). This kinase may not be present in mast cells or recruited by FcεRI.

FcγRIIB also become tyrosyl phosphorylated when coaggregated with c-Kit in mast cells. The phosphorylation of c-Kit was found not to be affected by coaggregation with FcγRIIB (MALBEC et al. 1999). Based on the role of FcεRI-associated kinases during FcγRIIB-FcεRI coaggregation, one can suggest that FcγRIIB might be phosphorylated by the kinase domain of c-Kit or by a nonreceptor PTK recruited by c-Kit upon c-Kit activation.

That phosphorylated FcγRIIB could recruit SH2-bearing phosphatases was first observed in B cells (D'AMBROSIO et al. 1995; ONO et al. 1996). Subsequent studies in mast cells showed that, when coaggregated with FcεRI, phosphorylated FcγRIIB recruit SHIP (ONO et al. 1996), but not SHP-1 or SHP-2 (FONG et al. 1996). FcγRIIB-mediated inhibition was conserved in mast cells (FONG et al. 1996; ONO et al. 1996), but not in B cells (D'AMBROSIO et al. 1995) derived from SHP-1-deficient moth-eaten mice. Supporting a predominant role of SHIP in FcγRIIB-mediated inhibition, including in B lymphocytes, chimeras made of the catalytic domain of SHIP in place of the FcγRIIB intracytoplasmic domain were able to reproduce the inhibition, when coaggregated with BCR in DT40 B cells (ONO et al. 1997). Moreover, FcγRIIB-mediated inhibition was conserved in SHP-1-deficient, but not in SHIP-deficient DT-40 cells (ONO et al. 1997). Finally, the coaggregation of FcεRI or BCR with FcγRIIB whose Y-2 isoleucine residue, in the ITIM motif, was mutated (this mutation was demonstrated to prevent FcγRIIB ITIM from associating with SHP-1) inhibited mast cell and B cell activation (Bruhns et al., submitted).

SHIP is an inositol polyphosphate 5-phosphatase (DAMEN et al. 1996), whose primary substrate is membrane-anchored phosphatidylinositol (3,4,5)-trisphosphate (PIP3). During cell activation, PIP3 is produced by the phosphorylation of PI(4,5)P2 by PI3 kinase (PI3K). PIP3 was described to interact with PH domains, particularly those of PTKs of the Tec family, such as Btk (SALIM et al. 1996). This interaction enables translocation of Btk to the plasma membrane, where it can interact with and be phosphorylated by lyn. Activated Btk then phosphorylates phospholipase C-γ (PLCγ). This enhances the production of inositol (1,4,5)-trisphosphate (IP3), which is responsible for the release of intracellular Ca^{2+} stores (CLAPHAM 1995). During coaggregation of FcγRIIB with FcεRI or with BCR, SHIP recruited by phosphorylated FcγRIIB can dephosphorylate PIP3 into PI(3,4)P2, thus preventing the membrane recruitment of Btk. Supporting this conclusion, the expression of Btk as a membrane chimera in DT-40 cells was sufficient to abrogate FcγRIIB-mediated inhibition, suggesting that the inhibition of recruitment and activation of Btk accounts for SHIP-mediated inhibition induced by FcγRIIB (BOLLAND et al. 1998). Btk activation was indeed shown to be required for the sustained production of IP3, initiated by PLC-γ, and for the subsequent opening of membrane Ca^{2+} channels (SCHARENBERG et al. 1998). As a consequence, FcγRIIB-mediated inhibition prevents the influx of extracellular calcium without affecting the initial mobilization of intracellular Ca^{2+} stores (CHOQUET et al. 1993; BLÉRY et al. 1997).

A comparable mechanism is likely to account for FcγRIIB-mediated inhibition of c-kit-dependent mast cell proliferation. Following coaggregation with c-Kit in mast cells, phosphorylated FcγRIIB were found to selectively recruit SHIP. The role of SHIP in the inhibition of c-Kit-mediated proliferation remains to be demonstrated. C-kit, however, activates PI3K which generates PIP3 (ROTTAPEL et al. 1991), the substrate of SHIP. Supporting the role of SHIP in FcγRIIB-mediated inhibition, several growth factors, including SCF, were found to induce exaggerated proliferative responses of progenitors of hematopoietic cells in SHIP-deficient mice (HELGASON et al. 1998).

The study of FcγRIIB-mediated inhibition mechanism disclosed a unique mode of cooperation between FcγRIIB and ITAM-bearing receptors. It documented the biological significance of the necessary coaggregation of inhibitory receptors with cell-activating receptors. By providing activated kinases, ITAM-bearing receptors indeed play an active role in their own inhibition by FcγRIIB. This mechanism differs in that way from mechanisms that account for inhibition of cell activation by all other ITIM-bearing receptors that recruit SHP-1 and/or SHP-2. KIRs, for example, which inhibit NK and T cell activation, are constitu-tively associated with the src family PTK lck in NK cells. KIRs are tyrosyl phos-phorylated when aggregated, and phosphorylation is further enhanced when they are coaggregated with ITAM-bearing receptors (VÉLY et al. 1997). Phosphorylated KIRs recruit SHP-1 (BURSHTYN et al. 1996) and SHP-2 (BRUHNS et al. 1999). Coaggregation thus brings KIR-associated protein tyrosine phosphatases in con-tact with their tyrosyl phosphorylated substrates. Upon coaggregation of FcγRIIIA with KIRs in T lymphocytes, the ITAMs of FcγRIIIA, and possibly lck, are dephosphorylated, and ZAP-70 and PLCγ are not activated (BINSTADT et al. 1996). In KIR-mediated inhibition, the inhibited receptors are therefore passive targets of phosphatases recruited by the inhibitory receptors, and initial steps of signal transduction are aborted. By contrast, FcγRIIB-mediated inhibition does not operate until signal transduction is initiated by ITAM-bearing receptors, and it acts downstream of ITAM phosphorylation and syk recruitment (MALBEC et al. 1998). The two mechanisms of inhibition may not have the same biological consequences on the physiology of the cell (DAËRON 1997b).

5 Biological Significance of FcγRIIB-dependent Negative Regulation

5.1 FcγRIIB in Development

T and B cell differentiation has been shown to depend on signals delivered by the TCR and BCR, respectively, when binding to MHC-associated peptides or antigens with variable affinities. The expression of a functional antigen receptor is indeed a requirement for both immature B lymphocytes to be positively selected and further differentiate (FANG et al. 1996), and for mature B cells to survive (RAJEWSKY 1997; LAM et al. 1997). Transgenic mice bearing a monoclonal BCR enabled the dem-onstration, on the other hand, that high-affinity interactions of lymphocytes with autoantigens can deliver apoptotic signals, thus providing a basis for negative selection (NEMAZEE and BÜRKI 1989). Because FcγRIIB are expressed early during ontogeny, one may propose that they might negatively regulate signals delivered to preB or preT cells when binding to antigen in the presence of appropriate anti-bodies. Whether negative signals delivered by FcγRIIB may affect either positive and/or negative selection is not known. That no gross perturbation in lymphocyte differentiation was noticed in FcγRIIB-deficient mice (TAKAI et al. 1996) does not preclude this possibility.

Our finding that FcγRIIB can negatively regulate c-kit-dependent mast cell proliferation, and probably the growth and differentiation of other cell types, induced by other RTKs, suggests that FcγRIIB may also contribute to the control of ontogeny of hematopoietic cells in general. C-kit, for example, is transiently expressed at precise times during B cell differentiation (on pre B-II cells, after the V-D rearrangement and before the D-J rearrangement) (OSMOND et al. 1998) when FcγRIIB have already been expressed. Whether FcγRIIB might control proliferative signals delivered by c-kit at these critical moments needs to be investigated. C-kit is also expressed by other hematopoietic cells, as well as by cells of melanocytic and germinal lineage. If many hematopoietic cells express FcγRIIB, it is not known which of them, among other c-kit-positive cells, also express FcγRIIB.

5.2 FcγRIIB During an Immune Response

Evidence that FcγRIIB-dependent negative regulation contributes to the regulation of antibody response was provided by the observation that the titer of serum antibodies was four times higher in FcγRIIB-deficient mice than in littermate controls, following immunization with T-dependent antigens, such as SRBC, TNP-KLH in Freund's complete adjuvant, or with T-independent antigens such as TNP-LPS or TNP-Ficoll (TAKAI et al. 1996). Surprisingly, however, the in vivo experimental model of passively injected, antibody-induced suppression which was at the origin of the concept may not be accounted for by FcγRIIB-dependent negative regulation. It was indeed recently reported that passively administered, anti-sheep red blood cell (SRBC) antibodies still suppressed anti-SRBC antibody responses in FcγRIIB-deficient mice (KARLSSON and HEYMAN 1997). These discrepant observations indicate that suppression by passive antibodies may not be a good model for the physiological negative feedback of antibody production during a normal immune response. They emphasize the fact that several mechanisms, other than FcγRIIB-dependent negative regulation, are likely to concur in regulation of such a complex process as antibody response (DAËRON and HEYMAN 1998).

That FcγRIIB might control cell-mediated immunity was suggested by the finding that coaggregation of TCR with FcγRIIB expressed on T cells inhibits the synthesis of IL-2 by T cells (DAËRON et al. 1995a). FcγRIIB-positive T helper cells might thus be prevented from being activated when recognizing MHC-associated peptides on antigen-presenting cells coated with IgG antibodies against any cell surface antigen. Likewise, FcγRIIB-positive cytotoxic T cells may be prevented from killing IgG antibody-coated target cells.

Finally, FcγRIIB expressed by mast cells and basophils may explain why 80% of individuals in developed countries do not suffer from allergic manifestations when exposed to the same environmental allergens as allergic patients. Anti-allergen IgG antibodies were indeed shown to be produced together with IgE antibodies by both normal individuals and allergic patients (AALBERSE et al. 1983). The high concentration of IgG antibodies in the extracellular medium makes it seem likely that many allergens are complexed with specific IgG antibodies before they come into contact with mast cells sensitized with IgE against that allergen. By coaggregating FcεRI and FcγRIIB, such immune complexes can be expected to maintain mast cell activation at a physiological level, below the threshold beyond which pathological symptoms develop. The constitutive coexpression of ITIM-bearing FcγRIIB and ITAM-bearing FcγRIIIA or FcγRIIA/C, by mouse mast cells (BENHAMOU et al. 1990) and human basophils (DAËRON et al. 1995a), respectively, explains several observations made in both mice and humans. In mice, peritoneal mast cells of the serosal type readily degranulate when exposed to IgG immune complexes (PROUVOST-DANON et al. 1966), whereas bone marrow-derived mast cells (BMMC) of the mucous type do not (BENHAMOU et al. 1990). We found that peritoneal mast cells express about five times more FcγRIIIA than FcγRIIB (our unpublished observation), whereas BMMC express primarily FcγRIIB and only traces of FcγRIIIA (DAËRON et al. 1995b). In vitro experiments using rat mast cells transfected with cDNA encoding FcγRIIIA (LATOUR et al. 1992; DAËRON et al. 1992) and in vivo experiments using FcγRIIIA-deficient mice (HAZENBOS et al. 1996) demonstrated that FcγRIIIA account for IgG-dependent mast cell degranulation and anaphylaxis. One can thus propose that, as a general rule, the biological response of cells that coexpress ITAM-bearing and ITIM-bearing receptors which bind the same extracellular ligands depends on the relative proportion of these receptors. Interestingly, the expression of FcγRIIIA (WEINSHANK et al. 1988) and of FcγRIIB (CONRAD et al. 1987) is differentially regulated. Supporting this proposal, it was found that, by contrast with wild-type BMMC, BMMC from FcγRIIB-deficient mice do degranulate in response to IgG immune complexes and that IgG-induced passive cutaneous anaphylactic reactions are augmented in FcγRIIB-deficient mice, compared to wild-type mice (TAKAI et al. 1996).

A similar control is likely to apply to other ITAM-bearing FcR-dependent effector functions performed by a variety of FcγRIIB-expressing cells. These include antibody-dependent cell-mediated cytotoxicity (ADCC) reactions performed by macrophages, monocytes or eosinophils against target cells coated with IgG, IgE, or IgA antibodies, as well as the release of mediators (e.g., enzymes, oxygen metabolites, lipid-derived mediators, and cytokines) by the same cells in response to antibodies of the same classes.

5.3 FcγRIIB as Potential Therapeutic Tools for Immunotherapy

The large array of biological functions likely to be under the control of FcγRIIB make these receptors potential tools for immunotherapy in a variety of diseases. Some empirical treatments may already use FcγRIIB; others may be designed for specific purposes.

The passive administration of IgG anti-D antibodies has proven to be an efficient prophylactic treatment of hemolytic disease in Rh-positive newborns from Rh-negative mothers (CLARKE 1967). Such antibodies may increase physiological negative feedback control of an ongoing anti-Rh antibody response. Whether the mechanism by which passive antibodies inhibit this response can be accounted for by FcγRIIB-mediated negative regulation has been suggested but not demonstrated.

Another therapeutic maneuver that may depend on the regulatory properties of FcγRIIB could be the remarkably efficient, but not yet understood, effects of immunoglobulins pooled from large numbers of normal donors when administered intravenously (IVIg) to autoimmune patients (KAZATCHKINE et al. 1994). One may indeed propose that natural autoantibodies to antigens borne by target cells of autoreactive T cells, or to extracellular domains of the TCR complex, may shut off cytotoxicity by coaggregating the TCR and FcγRIIB on these cells. If so, one might envision autoimmunity as the consequence of a defective or inefficient physiological humoral autoimmunity, rather than the result of exaggerated or inappropriate pathological cell-mediated autoimmunity.

Somehow less hypothetical is the role FcγRIIB might play in allergic patients submitted to desensitization. This immunotherapy is based on the injection of progressively increasing amounts of a specific allergen over long periods of time, and its efficacy generally correlates with the titers of allergen-specific IgG antibodies induced in the serum of these patients (GLEICH et al. 1978). Such a hyperimmunization may thus provide IgG antibodies that, for an unknown reason, may be lacking or inefficient in allergic patients. Desensitization, however, is, to say the least, not always efficient; and it would be interesting, from a prognostic point of view, to determine whether patients who respond positively to immunotherapy differ from patients who do not, in the ability of their mast cells to be negatively regulated by FcγRIIB.

Finally, the finding that FcγRIIB negatively regulate RTK-dependent cell proliferation may be exploited as an immunotherapeutic approach to some cancers. Most RTKs, indeed, exist in an oncogenic form, resulting from mutations which make them constitutively activated in the absence of ligands. The same point mutation was identified, for instance, in the kinase domain of c-kit expressed by mouse (TSUJIMURA et al. 1994), rat (TSUJIMURA et al. 1995), and human (FURITSU et al. 1993) mastocytoma cells; and it was demonstrated that the transfection of c-kit bearing this single amino acid substitution was sufficient to transform primary cells and render them growth factor-independent (KITAYAMA et al. 1996). Since we found that FcγRIIB inhibit c-kit-dependent cell proliferation by recruiting SHIP, the target of which stands downstream from receptor phosphorylation, one can propose that bispecific ligands capable of coaggregating c-kit (or another RTK) with FcγRIIB (such as, for instance, IgG antibodies to growth factor receptors or to their ligands) might be useful in stopping the proliferation of FcγRIIB-expressing malignant cells.

Acknowledgments. This work was supported by the Institut National de la Santé et de la Recherche Médicale (INSERM), the Institut Curie, and the Association pour la Recheche sur le Cancer (ARC).

References

Aalberse RC, van der Gaag R, van Leeuwen J (1983) Serologic aspects of IgG4 antibodies. I. Prolonged immunization results in an IgG4-restricted response. J Immunol 130:722–726

Amigorena S, Bonnerot C, Choquet D, Fridman WH, Teillaud JL (1989) FcγRII expression in resting and activated B lymphocytes. Eur J Immunol 19:1379–1385

Amigorena S, Bonnerot C, Drake J, Choquet D, Hunziker W, Guillet JG, Webster P, Sautès C, Mellman I, Fridman WH (1992) Cytoplasmic domain heterogeneity and functions of IgG Fc receptors in B-lymphocytes. Science 256:1808–1812

Avrameas S, Guilbert B, Dighiero G (1981) Natural antibodies against actin, tubulin, myoglobin, thyroglobulin, fetuin, albumin, and transferin are present in normal human sera and monoclonal immunoglobulins from multiple myeloma and Waldeström macroglobulinemia. Ann Immunol (Inst Pasteur) 132 C:231–240

Benhamou M, Bonnerot C, Fridman WH, Daëron M (1990) Molecular heterogeneity of murine mast cell Fcγ receptors. J Immunol 144:3071–3077

Binstadt BA, Brumbaugh KM, Dick CJ, Scharenberg AM, Williams BL, Colonna M, Lanier LL, Kinet J-P, Abraham RT, Leibson PJ (1996) Sequential involvement of lck and SHP-1 with MHC-recognizing receptors on NK cells inhibits FcR-initiated tyrosine kinase activation. Immunity 5: 629–638

Bléry M, Delon J, Trautmann A, Cambiaggi A, Olcese L, Biassoni R, Moretta L, Chavrier P, Moretta A, Daëron M, Vivier E (1997) Reconstituted killer-cell inhibitory receptors for MHC class I molecules control mast cell activation induced via immunoreceptor tyrosine-based activation motifs. J Biol Chem 272:8989–8996

Bolland S, Pearse RN, Kurosaki T, Ravetch JV (1998) SHIP modulates immune receptor responses by regulating membrane association of Btk. Immunity 8:509–516

Brooks DG, Qiu WQ, Luster AD, Ravetch JV (1989) Structure and expression of human IgG FcRII (CD32). Functional heterogeneity is encoded by the alternatively spliced products of multiple genes. J Exp Med 170:1369–1386

Bruhns P, Marchetti P, Fridman WH, Vivier E, Daëron M (1999) Differential roles of N- and C-terminal ITIMs during inhibition of cell activation by killer cell inhibitory receptors. J Immunol 162: 3168–3175

Burshtyn DN, Scharenberg AM, Wagtmann N, Rajogopalan S, Berrada K, Yi T, Kinet J-P, Long EO (1996) Recruitment of tyrosine phosphatase HCP by the killer cell inhibitory receptor. Immunity 4: 77–85

Burshtyn, DN, Yang W, Yi T, Long EO (1997) A novel phosphotyrosine motif with a critical amino acid at position-2 for the SH2 domain-mediated activation of the tyrosine phosphatase SHP-1. J Biol Chem 272:13066–13072

Cambier JC (1995) Antigen and Fc Receptor signaling: The awesome power of the immunoreceptor tyrosine-based activation motif (ITAM). J Immunol 155:3281–3285

Choquet D, Partiseti M, Amigorena S, Bonnerot C, Fridman WH, Korn H (1993) Cross-linking of IgG receptors inhibits membrane immunoglobulin-stimulated calcium influx in B lymphocytes. J Cell Biol 121:355–363

Clapham DE (1995) Calcium signaling. Cell 80:259–268

Clarke CA (1967) Prevention of Rh-haemolytic disease. Br Med J 4:7–12

Conrad DH, Waldschmidt TJ, Lee WT, Rao M, Keegan AD, Noelle RJ, Lynch RG, Kehry MR (1987) Effect of B cell stimulatory factor-1 (interleukin 4) of Fcε and Fcγ receptor expression on murine B lymphocytes and B cell lines. J Immunol 139:2290–2296

D'Ambrosio D, Hippen KH, Minskoff SA, Mellman I, Pani G, Siminovitch KA, Cambier JC (1995) Recruitment and activation of PTP1 C in negative regulation of antigen receptor signaling by FcγRIIB1. Science 268:293–296

Daëron M (1997a) Fc Receptor Biology. Annu Rev Immunol 15:203–234

Daëron M (1997b) ITIM-bearing negative coreceptors. The Immunologist 5:79–86

Daëron M, Bonnerot C, Latour S, Fridman WH (1992) Murine recombinant FcγRIII, but not FcγRII, trigger serotonin release in rat basophilic leukemia cells. J Immunol 149:1365–1373

Daëron M, Heyman B (1998) FcgR and IgG-mediated negative regulation of immune responses. The immunoglobulin receptors and their physiological and pathological roles in immunity, J. van de Winkel and PM. Hogarth Eds, Kluwer Academic Publishers, pp 155–167

Daëron M, Latour S, Malbec O, Espinosa E, Pina P, Pasmans S, Fridman WH (1995a) The same tyrosine-based inhibition motif, in the intracytoplasmic domain of FcγRIIB, regulates negatively BCR-, TCR-, and FcR-dependent cell activation. Immunity 3:635–646

Daëron M, Malbec O, Latour S, Arock M, Fridman WH (1995b) Regulation of high-affinity IgE receptor-mediated mast cell activation by murine low-affinity IgG receptors. J Clin Invest 95: 577–585

Daëron M, Malbec O, Latour S, Bonnerot S, Segal DM, Fridman WH (1993) Distinct intracytoplasmic sequences are required for endocytosis and phagocytosis via murine FcγRII in mast cells. Intern Immunol 5:1393–1401

Damen JE, Liu L, Rosten P, Humphries RK, Jefferson AB, Majerus PW, Krystal G (1996) The 145-kDa protein induced to associate with Shc by multiple cytokines is an inositol tetraphosphate and phosphatidylinositol 3,4,5-trisphosphate 5-phosphatase. Proc Natl Acad Sci USA 93:1689–1693

Dubreuil P, Rottapel RR, AD, Forrester L, Bernstein A (1990) The mouse W/c-kit locus. A mammalian gene that controls the development of three distinct cell lineages. Ann NY Acad Sci 599:58–65

Esposito-Farese M-E, Sautès C, de la Salle H, Latour S, Bieber T, de la Salle C, Ohlmann P, Fridman WH, Cazenave J-P, Teillaud J-L, Daëron M, Bonnerot C, Hanau D (1995) Membrane and soluble FcγRII/III modulate the antigen-presenting capacity of murine dendritic epidermal Langerhans cells for IgG-complexed antigens. J Immunol 154:1725–1736

Fang W, Mueller DL, Pennell CA, Rivard JJ, Li YS, Hardy RR, Schlissel MS, Behrens TW (1996) Frequent aberrant, immunoglobulin gene rearrangements in pro-B cells revealed by a bcl-xL transgene. Immunity 4:291–299

Fong DC, Malbec O, Arock M, Cambier JC, Fridman WH, Daëron M (1996) Selective in vivo recruitment of the phosphatidylinositol phosphatase SHIP by phosphorylated FcγRIIB during negative regulation of IgE-dependent mouse mast cell activation. Immunol Letters 54:83–91

Furitsu T, Tsujimura T, Tono T, Ikeda H, Kitayama H, Koshimizu U, Sugahara H, Butterfield JH, Ashman LK, Kanayama Y, Matsuzawa Y, Kitamura Y, Kanakura Y (1993) Identification of mutations in the coding sequence of the proto-oncogene c-kit in a human mast cell leukemia cell line causing ligand-independent activation of c-kit product. J Clin Invest 92:1736–1744

Gleich GJ, Zimmermann EM, Henderson LL, Yunginger JW (1978) Effect of immunotherapy on immunoglobulin E and immunoglobulin G antibodies to ragweed antigens: a six-year prospective study. J Allergy Clin Immunol 62:261   ·

Hazenbos LW, Gessner JE, Hofhuis FMA, Kuipers H, Meyer D, Heijnen IAFM, Schmidt RE, Sandor M, Capel PJA, Daëron M, van de Winkel JGJ, Verbeek JS (1996) Impaired IgG-dependent anaphylaxis and Arthus reaction in FcγRIII (CD16) deficient mice. Immunity 5:181–188

Helgason CD, Damen JE, Rosten P, Grewal R, Sorensen P, Chappel SM, Borowski A, Jirik F, Krystal G, Humphries RK (1998) Targeted disruption of SHIP leads to hemopoietic perturbations, lung pathology, and shortened life span. Genes Dev 12:1610–1620

Henry C, Jerne NK (1968) Competition of 19S and 7S antigen receptors in the regulation of the primary immune response. J Exp Med 128:133–152

Heyman B (1989) Inhibition of IgG-mediated immunosuppression by a monoclonal anti-Fc-receptor antibody. Scand J Immunol 29:121–126

Heyman B (1990) Fc-dependent IgG-mediated suppression of the antibody response: fact or artefact? Scand J Immunol 31:601–607

Hibbs ML, Walker ID, Kirszbaum L, Pietersz GA, Deacon NJ, Chambers GW, McKenzie IFC, Hogarth PM (1986) The murine Fc receptor for immunoglobulin: purification, partial amino acid sequence, and isolation of cDNA clones. Proc Natl Acad Sci USA 83:6980–6984

Hulett MD, Hogarth PM (1994) Molecular basis of Fc receptor function. Adv Immunol 57:1–127

Karlsson M, Heyman B (1997) Unimpaired feedback suppression by IgG in FcγRIIB-deficient mice. Immunol Let 56:205

Kazatchkine M, Dietrich G, Hurez V, Ronda N, Bellon B, Rossi F, Kaveri SV (1994) V region-mediated selection of autoreactive repertoires by intravenous immunoglobulin (i.v.Ig). Immunol Rev 139: 79–107

Kitayama H, Tsujimura T, Matsumara I, Oritani K, Ikeda H, Ishikawa J, Okabe M, Suzuki M, Yamamura K-I, Matsuzawa Y, Kitamura Y, Kanakura Y (1996) Neoplastic transformation of normal hematopoietic cells by constitutively activated mutations of c-kit receptor tyrosine kinase. Blood 88:995–1004

Kohler H, Richardson B, Rowley DA, Smyk S (1977) Immune response to phosphorylcholine. III. Requirement of the Fc portion and equal effectiveness of IgG subclasses in anti-receptor antibody-induced suppression. J Immunol 119:1979–1986

Kohler H, Richardson BC, Smyk S (1978) Immune response to phosphorylcholine. IV. Comparison of homologous and isologous anti-idiotypic antibody. J Immunol 120:233–238

Lam KP, Kuhn R, Rajewski K (1997) In vivo ablation of surface immunoglobulin on mature B cells by inducible gene targeting results in rapid cell death. Cell 90:1073–1083

Latour S, Bonnerot C, Fridman WH, Daëron M (1992) Induction of tumor necrosis factor-α production by mast cells via FcγR. Role of the FcγRIIIγ subunit. J Immunol 149:2155–2162

Latour S, Fridman WH, Daëron M (1996) Identification, molecular cloning, biological properties, and tissue distribution of a novel isoform of murine low-affinity IgG receptor homologous to human FcγRIIB1. J Immunol 157:189–197

Lewis VA, Koch T, Plutner H, Mellman I (1986) A complementary DNA clone for a macrophage-lymphocyte Fc receptor. Nature 324:372

Long EO, Burshtyn DN, Clark WP, Peruzzi M, Rajagopalan S, Rojo S, Wagtmann N, Winter CC (1997) Killer cell inhibitory receptors: diversity, specificity, and function. [Review] [51 refs]. Immunol Rev 155:135–144

Lynch RG, Sandor M (1990) Fc receptors on T and B lymphocytes. Fc receptors and the action of antibodies 13, H. Metzgzer Ed, pp. 305–334

Malbec O, Fong D, Turner M, Tybulewicz VLJ, Cambier JC, Fridman WH, Daëron M (1998) FcεRI-associated lyn-dependent phosphorylation of FcγRIIB during negative regulation of mast cell activation. J Immunol 160:1647–1658

Malbec O, Fridman WH, Daëron M (1999) Negative regulation of c-kit-mediated cell proliferation by FcγRIIB. J Immunol 162:4424–4429

Manning DD, Jutila JW (1972) Immunosuppression of mice injected with heterologous anti-immunoglobulin heavy chain antisera. J Exp Med 135:1316–1323

Mantzioris BX, Berger MF, Sewell W, Zola H (1993) Expression of the Fc receptor for IgG (FcγRII/CDw32) by human circulating T and B lymphocytes. J Immunol 150:5175–5184

Mellman I, Koch T, Healey G, Hunziker W, Lewis V, Plutner H, Miettinen H, Vaux D, Moore K, Stuart S (1988) Structure and function of Fc receptors on macrophages and lymphocytes. [Review]. Journal of Cell Science Supplement 9:45–65

Muta T, Kurosaki T, Misulovin Z, Sanchez M, Nussenzweig MC, Ravetch JV (1994) A 13-amino-acid motif in the cytoplasmic domain of FcγRIIB modulates B-cell receptor signalling. Nature 368:70–73

Nemazee DA, Bürki K (1989) Clonal deletion of B lymphocytes in a transgenic mouse bearing anti-MHC class I antibody genes. Nature 337:562–566

Ogawa M, Matsuzaki Y, Nishikawa S, Hayashi S, Kunisada T, Sudo T, Kina T, Nakauchi H, Nishikawa S (1991) Expression and function of c-kit in hematopoietic cells. J Exp Med 174:63–71

Olcese L, Lang P, Vély F, Cambiaggi A, Marguet D, Bléry M, Hippen KL, Biassoni R, Moretta A, Moretta L, Cambier JC, Vivier E (1996) Human and mouse killer cell inhibitory receptors recruit PTP1 C and PTP1D protein tyrosine phosphatases. J Immunol 156:4531–4534

Ono M, Bolland S, Tempst P, Ravetch JV (1996) Role of the inositol phosphatase SHIP in negative regulation of the immune system by the receptor FcγRIIB. Nature 383:263–266

Ono M, Okada H, Bolland S, Yanagi S, Kurosaki T, Ravetch JV (1997) Deletion of SHIP or SHP-1 reveals two distinct pathways for inhibitory signaling. Cell 90:293–301

Osmond DG, Rolink A, Melchers F (1998) Murine B lymphopoiesis: towards a unified model. Immunol Today 19:65–68

Phillips NE, Parker DC (1984) Cross-linking of B lymphocyte Fcγ receptors and membrane immunoglobulin inhibits anti-immunoglobulin-induced blastogenesis. J Immunol 132:627–632

Pleinman C, D'Ambrosio D, Cambier J (1994) The B-cell antigen receptor complex: structure and signal transduction. Immunol Today 15:393–399

Prouvost-Danon A, Queiroz-Javierre M, Silva-Lima M (1966) Passive anaphylactic reaction in mouse peritoneal mast cells in vitro. Life Sci 5:1751

Rajewsky K (1997) Clonal selection and learning in the antibody system. Nature 381:751–758

Ravetch JV, Luster AD, Weinshank R, Kochan J, Pavlovec A, Portnoy DA, Hulmes J, Pan YCE, Unkeless JC (1986) Structural heterogeneity and functional domains of murine Immunoglobulin G Fc receptors. Science 234:718–725

Reth MG (1989) Antigen receptor tail clue. Nature 338:383–384

Rottapel R, Reedijk M, Williams DE, Lyman SD, Anderson DM, Powson T, Bernstein A (1991) The Steel/W transduction pathway: kit autophosphorylation and its association with a unique subset of cytoplasmic signaling proteins is induced by steel factor. Mol Cell Biol 11:3043–3051

Salim K, Bottomley MJ, Querfurth E, Zvelebil MJ, Gout I, Scaife R, Margolis RL, Gigg R, Smith CI, Driscoll PC, Waterfield MD, Panayotou G (1996) Distinct specificity in the recognition of phosphoinositides by the pleckstrin homology domains of dynamin and Bruton's tyrosine kinase. EMBO J 15:6241–6250

Sandor M, Lynch RG (1992) Lymphocyte Fc receptors: the special case of T cells. Immunol Today 14:227–231

Scharenberg AM, El-Hillal O, Fruman DA, Beitz LO, Li Z, Lin S, Gout I, Cantley LC, Rawlings DJ, Kinet J-P (1998) Phosphatidylinositol-3,4,5-triphosphate (PtdIns-3,4,5-P3)/Tec kinase-dependent calcium signaling pathway: a target for SHIP-mediated inhibitory signals. EMBO J 17:1961–1972

Scharenberg AM, Kinet J-P (1996) The emerging field of receptor-mediated inhibitory signaling: SHP or SHIP? Cell 87:961–964

Scharenberg AM, Lin S, Cuénod B, Yamamura H, Kinet J-P (1995) Reconstitution of interactions between tyrosine kinases and the high affinity IgE receptor which are controlled by receptor clustering. EMBO J 14:3385–3394

Shek PN, Dubiski S (1975) Allotypic suppression in rabbits: competition for target cell receptors between isologous and heterologous antibody, and between native antibody and antibody fragments. J Immunol 114:621

Sinclair NRSC, Chan, PL (1971) Regulation of the immune responses. IV. The role of the Fc-fragment in feedback inhibition by antibody. Adv Exp Med Biol 12:609–615

Takai T, Ono M, Hikida M, Ohmori H, Ravetch JV (1996) Augmented humoral and anaphylactic responses in FcγRII-deficient mice. Nature 379:346–349

Tsujimura T, Furitsu T, Morimoto M, Isozaki K, Nomura S, Matsuzawa Y, Kitamura Y, Kanakura Y (1994) Ligand-independent activation of c-kit receptor tyrosine kinase in a murine mastocytoma cell line P-815 generated by a point mutation. Blood 83:2619–2626

Tsujimura T, Furitsu T, Morimoto M, Kanayama Y, Nomura S, Matsuzawa Y, Kitamura Y, Kanakura Y (1995) Substitution of an aspartic acid results in constitutive activation of c-kit receptor tyrosine kinase in rat tumor mast cell line RBL-2H3. Int Arch Allergy Immunol 106:377–385

Unkeless JC, Jin J (1997) Inhibitory receptors, ITIM sequences, and phosphatases. Curr Opin Immunol 9:338–343

Vély F, Olivero S, Olcese L, Moretta A, Damen JE, Liu L, Krystal G, Cambier JC, Daëron M, Vivier E (1997) Differential association of phosphatases with hematopoietic coreceptors bearing immunoreceptor tyrosine-based inhibition motifs. Eur J Immunol 27:1994–2000

Vivier E, Daëron M (1997) Immunoreceptor tyrosine-based inhibition motifs. Immunol Today 18: 286–291

Wang J, Koizumi T, Watanabe T (1996) Altered antigen receptor signaling and impaired Fas-mediated apoptosis of B cells from lyn-deficient mice. J Exp Med 184:831

Weinshank RL, Luster AD, Ravetch JV (1988) Function and regulation of the murine-macrophage-specific IgG Fc receptor, FcγR-α. J Exp Med 167:1909–1925

The Role of the SRC Homology 2-Containing Inositol 5'-Phosphatase in FcεR1-Induced Signaling

M. Huber[1], C.D. Helgason[1], J.E. Damen[1], M.P. Scheid[2], V. Duronio[2],
V. Lam[1], R.K. Humphries[1], and G. Krystal[1]

1 Introduction

In 1996, three groups independently cloned the hemopoietic specific src homology 2 (SH2)-containing inositol 5'-phosphatase, SHIP (Damen et al. 1996; Lioubin et al. 1996; Kavanaugh et al. 1996). Although this intracellular enzyme, which is capable in vitro of hydrolyzing the 5'-phosphate from phosphatidylinositol-3,4,5-trisphosphate (PIP_3) and inositol-1,3,4,5-tetrakisphosphate (IP_4) (Damen et al. 1996), was originally identified as a 145-kDa protein that became both tyrosine phosphorylated and associated with the adaptor protein Shc in response to multiple cytokines and to B- and T-cell receptor engagement (reviewed in Liu et al. 1997), it is currently in the limelight because of its interaction with the negative co-receptor FcγRIIB. Specifically, SHIP is now known to inhibit immune receptor-mediated activation in both mast cells and B cells by binding to the tyrosine phosphorylated immunoreceptor tyrosine-based inhibitory motif (ITIM) of the inhibitory coreceptor FcγRIIB (Vely et al. 1997; Tridandapani et al. 1997) and inhibiting FcεR1- and B cell receptor-induced degranulation and proliferation, respectively (Ono et al. 1996; Ono et al. 1997). However, while activation of the B cell receptor in the absence of FcγRIIB co-clustering does not lead to significant tyrosine phosphorylation of SHIP (Chacko et al. 1996), activation of the FcεR1 alone does

[1] The Terry Fox Laboratory, BC Cancer Agency, Vancouver, BC, Canada V5Z 1L3
[2] The Jack Bell Research Centre, Vancouver, BC, Canada V6H 3Z6

(HUBER et al. 1998). We therefore investigated whether SHIP plays any role in modulating IgE-induced signaling via this immunoreceptor tyrosine based activation motif (ITAM)-containing receptor in the absence of FcγRIIB co-clustering. In this review we discuss our findings in this area.

2 The Characterization of SHIP

SHIP possesses an amino terminal SH2 domain, two centrally located motifs highly conserved among inositol polyphosphate 5'-phosphatases (5-ptases), two NPXY sequences that, when phosphorylated, can bind phosphotyrosine binding (PTB) domains, and a proline rich C-terminus theoretically capable of binding to many SH3-containing proteins (reviewed in LIU et al. 1997) (see Fig. 1). Interestingly, unlike most 5-ptases, SHIP selectively hydrolyzes, at least in vitro, the 5'-phosphate from PIP_3 and IP_4 (DAMEN et al. 1996), two phosphoinositides that have been shown to play important roles in growth factor-mediated signaling (IRVINE 1998). The human and murine forms of SHIP possess 87.2% identity at the amino acid level and human SHIP maps to the long arm of chromosome 2 at the border between 2q36 and 2q37 (WARE et al. 1996). During murine development, SHIP is first detectable, by restriction transciptase polymerase chain reaction (RT-PCR), in 7.5 day postcoitus mouse embryos (coincident with and dependent upon the onset of hemopoiesis), and its protein expression pattern in the embryo appears restricted to hemopoietic cells (LIU et al. 1998). In the adult mouse, SHIP protein expression is also restricted to hemopoietic cells (and to spermatids) (LIU et al. 1998). Also worthy of note is that SHIP protein expression appears to vary considerably during hemopoiesis (GEIER et al. 1997; LIU et al. 1998), increasing substantially, for example, with T cell maturation (LIU et al. 1998).

Since SHIP's 5-ptase activity does not change significantly following cytokine stimulation (DAMEN et al. 1996), it likely exerts its downstream effects via intracellular translocation, perhaps via association with other proteins. We have therefore sought to identify binding partners in different hemopoietic cell types. These studies have shown that SHIP associates with Shc (via the SH2 and NPXYs

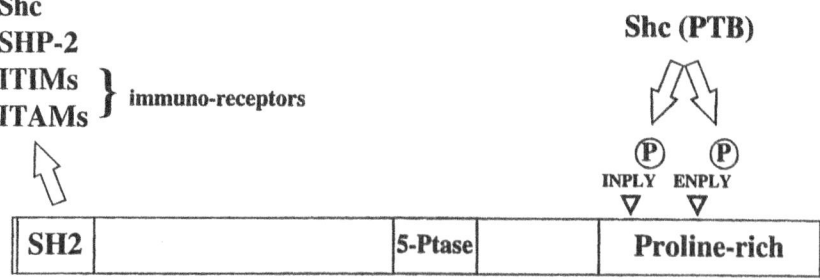

Fig. 1. The structure of SHIP and its currently known binding partners

of SHIP with the $pY^{317}VNV$ and PTB motifs of Shc, respectively, in IL-3 stimulated myeloid cell lines (LIU et al. 1997a) and via the NPXYs of SHIP with the PTB domain of Shc in T cell receptor-activated T cells (LAMKIN et al. 1997)). SHIP also associates with the tyrosine phosphatase, SHP-2 (via the SH2 domain of SHIP with the $pY^{542}XNI$ or $pY^{580}XNV$ of SHP-2 (LIU et al. 1997b)). This latter finding has been demonstrated independently by SATTLER et al. (1997) and it is possible that SHP-2 binding leads to the dephosphorylation of SHIP (LIU et al. 1997b). SHIP has also been shown to bind via its SH2 domain to the ITIM of certain members of the ITIM-bearing family of inhibitory co-receptors. Specifically, it has been shown to bind both in vitro (VELY et al. 1997; TRIDANDAPANI et al. 1997) and in vivo (ONO et al. 1997) to the ITIM of FcγRIIB, and in vitro to the second ITIM of gp49B1 (KUROIWA et al. 1998), but not to either ITIM of the killer cell inhibitory receptor (KIR) (VELY et al. 1997)). Furthermore, SHIP's SH2 domain, which binds preferentially to the sequence $pY(Y/D)X(L/I/V)$ (OSBORNE et al. 1996), has been demonstrated to bind in vitro to the tyrosine phosphorylated ITAM within the β (KIMURA et al. 1997) and γ (OSBORNE et al. 1996) subunits of the FcεR1 as well as the ζ chain of the T cell receptor (OSBORNE et al. 1996). However, we have been unsuccessful in attempts to demonstrate any in vivo association, via co-precipitation studies with various detergents and buffers, between SHIP and the FcεR1 (Huber et al. unpublished).

Complicating matters, SHIP exists in at least four molecular weight forms; and we have evidence that the lower 135-, 125-, and 110-kDa forms are generated from the 145-kDa full length protein in vivo by cleavage of its proline-rich C-terminus (DAMEN et al. 1998). While all forms become tyrosine phosphorylated in response to cytokines, at one or both of the NPXY motifs, only the 145- and, to a lesser extent, the 135-kDa species bind Shc; and only the 110-kDa form is associated with the cytoskeleton (DAMEN et al. 1998). Other groups have also reported the presence of multiple forms of SHIP (KAVANAUGH et al. 1996; ODAI et al. 1997) and shown that the relative proportion of the different forms changes with hemopoietic differentiation (GEIER et al. 1997) and with leukemogenesis (ODAI et al. 1997; SATTLER et al. 1997). Our results, however, do not rule out the possibility that additional SHIP proteins result from alternate splicing and, in fact, minor 6-kb, 4.5-kb, and smaller SHIP mRNA species have been reported (DAMEN et al. 1996; WARE et al. 1996; GEIER et al. 1997).

3 The Generation of a SHIP Knockout Mouse

To gain further insight into the role that SHIP plays in vivo, we recently generated a SHIP knockout mouse by homologous recombination in embryonic stem (ES) cells (HELGASON et al. 1998). Based on our results, showing that the multiple forms of SHIP were generated by C-terminal truncation of the protein product of a single mRNA (DAMEN et al. 1998), we felt confident targeting the first exon (the

SH2-containing region) of the SHIP gene. Hemopoietic cells from these mice lacked all four SHIP bands while their normal littermates possessed all four, further establishing that these proteins were derived from the same gene. Although these mice are viable and fertile, they suffer from progressive splenomegaly, massive macrophage infiltration of the lungs, wasting and a shortened lifespan (HELGASON et al. 1998). Notably, granulocyte/macrophage progenitors from the bone marrow of these mice are substantially more responsive to multiple cytokines (including interleukin-3, granulocyte-macrophage-colony stimulating factor and Steel factor) than their wild-type littermates (HELGASON et al. 1998), suggesting that SHIP is a negative regulator of hemopoietic growth factor induced proliferation/survival. Even in the absence of added growth factors, small colonies (i.e., less than 20 cells each) are present at day 10 in 15% FCS containing methylcellulose cultures from SHIP−/− but not +/+ bone marrow, suggesting once again that SHIP−/− progenitors have a survival advantage. It is likely that the phenotype of the SHIP−/− mouse is due in large part to the hyper-responsiveness of granulocyte/macrophage progenitors since normal mice transplanted with bone marrow cells engineered for retroviral-mediated overexpression of GM-CSF have a very similar phenotype (i.e., splenomegaly, reductions in bone marrow cellularity and patchy consolidation of the lungs (JOHNSON et al. 1989). Intriguingly, unlike granulocyte–macrophage progenitors, late erythroid (CFU-E) and B cell progenitors are reduced in SHIP−/− mice (HELGASON et al. 1998), and it is possible that SHIP plays an opposing role in these lineages. Alternatively, this reduction may not be an intrinsic property of these progenitors but rather the result of a displacement or inhibition by the increased numbers of granulocyte–macrophage progenitors and this is currently under investigation.

4 The Role of SHIP in FcεR1-induced Signaling

The FcεR1 (i.e., the high-affinity receptor for IgE) is composed of an α-, a β- and two covalently linked γ-subunits (ALBER et al. 1991). The cytoplasmic domains of the β and γ subunits contain ITAMs that are critical for IgE-mediated activation (ALBER et al. 1991). Since the SH2 domain of SHIP was shown recently to be capable of binding to these tyrosine phosphorylated ITAMs in vitro (KIMURA et al. 1997; OSBORNE et al. 1996), we investigated the role of SHIP in FcεR1 activation using primary mast cells derived from SHIP+/+ and −/− mice. BMMCs were prepared from 4–8 week F2 littermates, and by 8 weeks in culture, greater than 99% of the cells in the two cultures were both c-kit and FcεR1 positive (with similar mean fluorescences) (HUBER et al. 1998). Since the presence of these two cell surface proteins is a hallmark of mature mast cells, it indicated that SHIP was not essential for the generation and differentiation of BMMCs.

The conventional assay for measuring the IgE-induced degranulation of mast cells involves a preliminary passive presensitization step in which the cells are exposed to an IgE which has been generated against a specific antigen [e.g., a dinitrophenyl (DNP) group]. For this step the cells are either exposed to a vast excess of IgE or incubated overnight because of the very slow association rate constant for this interaction (i.e., $3–10 \times 10^4 M^{-1} \cdot s^{-1}$) (MACGLASHAN et al. 1998). Once a substantial number of the 100,000 to 300,000 FcεR1/cell have bound IgE, the antigen, in a multivalent form [e.g., 20–30moles of DNP covalently attached to human serum albumin (DNP-HSA)] is then added to induce FcεR1 aggregation. This latter, activating step is thought to initiate a biochemical cascade that results in degranulation and the release of mediators of allergic reactions (ALBER et al. 1991).

To gain some insight into the role of SHIP in IgE-mediated degranulation, we compared SHIP + / + and −/− BMMCs for their ability to degranulate using this conventional assay and found that the SHIP−/− cells consistently released substantially more of their total granule contents than the + / + cells (HUBER et al. 1998). Even more remarkable, however, was the finding that when IgE alone was added (i.e., the presensitization step), there was a massive degranulation of SHIP−/− BMMCs, whereas SHIP + / + cells, as previously reported, did not degranulate at all (HUBER et al. 1998). This suggested that SHIP plays an important role in preventing degranulation from occurring in the absence of antigen.

Since FcγRIIB co-clustering studies in mast cells had demonstrated earlier that SHIP inhibited FcεR1 activation by inhibiting extracellular calcium entry (ONO et al. 1996, 1997), we also investigated the relationship between the presence of SHIP and the entry of extracellular calcium in our system. We found that when SHIP + / + and −/− BMMCs were loaded with fura-2/AM, and either treated with IgE alone, or sensitized with IgE-anti-DNP and stimulated with DNP-HSA, the peak of intracellular calcium was substantially higher in the SHIP−/− BMMCs, consistent with their ability to degranulate. In all cases, addition of EGTA to deplete extracellular calcium both prevented degranulation and markedly reduced the increase in intracellular calcium to a level that was indistinguishable in + / + and −/− mast cells. This suggested that the increase in intracellular calcium in response to either IgE alone or IgE plus DNP-HSA was predominantly from the extracellular medium and that SHIP was acting most likely downstream of the initial release of intracellular calcium stores.

Although current dogma states that crosslinking of the IgE-preloaded FcεR1 by a multivalent antigen is the essential first step in triggering the signaling cascades that lead to degranulation, both the calcium influx into both SHIP + / + and −/− BMMCs and the massive degranulation in SHIP−/− BMMCs with IgE alone suggested that substantial intracellular signaling is already occurring during the IgE preloading step (i.e., in the absence of antigen). Interestingly, when we purified monomeric IgE from the commercial preparations of IgE-anti-DNP (IgE mAb SPE-7 from Sigma, which contained approx. 5%–10% aggregates) using BioSep SEC S3000 (Phenomenex) HPLC chromatography (which separates monomeric IgE from dimers and larger aggregates on the basis of size), we still obtained degranulation of SHIP−/− BMMCs. Moreover, rechromatography of the monomeric IgE anti-DNP, following its incubation at 37°C for 1h in Tyrode's buffer (an incubation period substantially longer than any of our assays) demonstrated that this monomeric IgE-stimulated degranulation was not due to aggregation during the assay period (Fig. 2). However, although no detectable aggregates were observed, we cannot rule out that a very low level of aggregation of IgE, either while in solution (since our lower limit of detection of IgE aggregates following HPLC is 0.5% of the total protein eluted) or following binding to cell surface FcεR1 is responsible for the observed IgE-induced degranulation of SHIP−/− BMMCs. Regardless, during this passive presensitization step, our degranulation results suggest that SHIP acts as a "gatekeeper", preventing the IgE-induced degranulation signal from progressing unless overwhelmed by the massive signaling initiated via crosslinking agents. To test this hypothesis, we explored tyrosine phosphorylation events elicited by IgE alone in SHIP+/+ and −/− BMMCs and found that both the β- and γ-subunits of the FcεR1 are indeed tyrosine phosphorylated in both cell types in response to IgE alone. Moreover, in +/+ cells, SHIP is constitutively tyrosine phosphorylated and its phosphorylation level increases upon addition of IgE. Intriguingly, immunoprecipitating Shc from these cells revealed that its tyrosine phosphorylation also occurs with IgE stimulation alone but its phosphorylation is markedly reduced in SHIP−/− BMMCs. This suggests that Shc needs to bind to SHIP in order to become tyrosine phosphorylated after FcεR1 activation by IgE (HUBER et al. 1998). The same result was obtained after antigen crosslinking of IgE-preloaded cells.

The mitogen-activated protein kinases (MAPKs) ERK 1 and 2 are also tyrosine phosphorylated by IgE alone in SHIP+/+ and −/− BMMCs, but in SHIP −/− cells this phosphorylation is greater and dramatically prolonged. This is interesting because this prolonged activation of ERK in SHIP−/− cells occurs under conditions where the tyrosine phosphorylation of Shc is markedly reduced, suggesting that Shc-dependent activation of the Ras pathway may not play a significant role in the IgE-mediated activation of MAPK, at least in SHIP−/− BMMCs. Interestingly, EGTA significantly reduced IgE-induced Erk 1 and 2 phosphorylation, especially in SHIP−/− cells, suggesting that the entry of extracellular calcium is upstream of MAPK activation in these cells (HUBER et al. 1998).

Since the 5-ptase activity of SHIP appears to be critical for reducing calcium entry during FcγRIIB mediated inhibition (ONO et al. 1997), we investigated whether it was SHIP's ability to hydrolyze PIP_3 or IP_4 This was important in restricting IgE-mediated extracellular calcium entry by testing the effects of PI-3-kinase inhibitors. In preliminary experiments we established that both wortmannin and LY294002 completely inhibited degranulation in SHIP−/− cells induced by IgE alone. This is consistent with previous results obtained with IgE-crosslinked RBL-2H3 cells (YANO et al. 1993). We then explored the effects of these inhibitors on the IgE-mediated influx of extracellular calcium. Dose-response studies revealed that as little as 10μM LY294002 reduced extracellular calcium influx into SHIP−/− cells to levels observed in IgE-stimulated SHIP+/+ BMMCs. On the other hand, LY294002 had far less effect on IgE-stimulated calcium influx in SHIP+/+ cells, even at 100μM concentrations. Similar results were obtained with wortmannin. These results suggest that during incubation of SHIP+/+ BMMCs with IgE, activation of PI-3 kinase and generation of PIP_3 plays a less significant

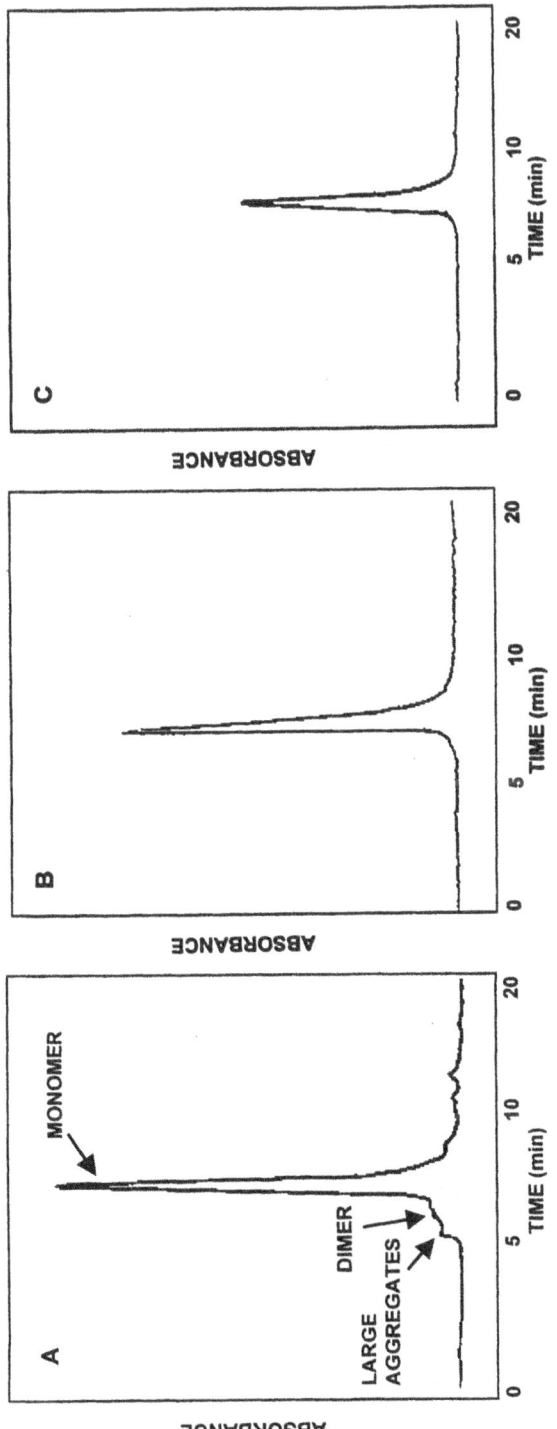

Fig. 2A–C Monomeric IgE does not aggregate during intracellular calcium or degranulation assay procedures. **A** Monomeric IgE was purified, using BioSep SEC S3000 HPLC (Phenomenex, Belmont, CA), and rechromatographed **B** immediately or **C** after 1-h incubation at 37°C in Tyrode's buffer. The larger aggregates shown in A elute in the void volume

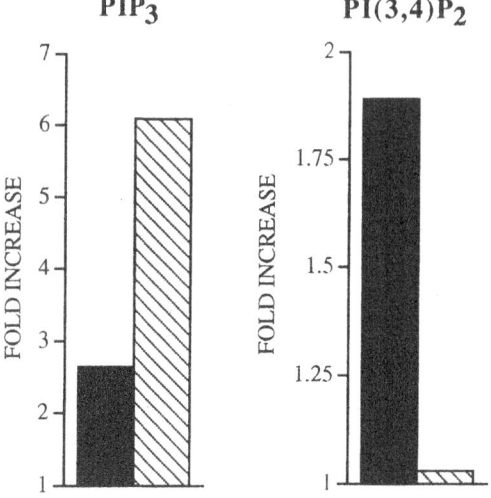

Fig. 3. PIP_3 and $PI\text{-}3,4\text{-}P_2$ levels in SHIP$-/-$ and $+/+$ BMMCs following 2min of stimulation with IgE. The cells were labeled for 90min with ^{32}P-ortho-phosphate, treated for 2min $+/-$ 10μg/ml IgE, and the labeled phosphoinositides extracted, deacylated, subjected to ion exchange chromatography using a Partisil 10 SAX column, the peaks corresponding to the deacylated PIP_3 and $PI\text{-}3,4\text{-}P_2$ counted in a liquid scintillation counter as described by Scheid and Duronio (1996). The *solid back bars* depict the values obtained from the SHIP$+/+$ and the hatched bars from the $-/-$ BMMCs

role in mediating extracellular calcium entry because SHIP effectively hydrolyses this phospholipid to $PI\text{-}3,4\text{-}P_2$. Thus, degranulation in the absence of antigen is prevented. However, in the absence of SHIP, IgE alone can increase PIP_3 to levels that augment extracellular calcium entry sufficiently for degranulation.

To confirm this, we measured PIP_3 and $PI\text{-}3,4\text{-}P_2$ levels in SHIP$+/+$ and $-/-$ cells in response to IgE alone. As predicted, PIP_3 levels increased far higher, and $PI\text{-}3,4\text{-}P_2$ levels far lower, following 2min of IgE-stimulation of SHIP$-/-$ BMMCs (Fig 3). This demonstrates that SHIP is the primary enzyme responsible for breaking down PIP_3 in vivo in response to IgE-induced activation of PI-3-kinase. Of note, PI-3-kinase activity in anti-phosphotyrosine immunoprecipitates from IgE-stimulated SHIP$+/+$ and $-/-$ BMMCs was identical, demonstrating that the elevation of PIP_3 levels in the $-/-$ cells was not due to a difference in PI-3-kinase activity in the two cell types.

Taken together, our results suggest a model (Fig 4) in which binding of IgE alone to the FcεR1 on normal primary mast cells activates an associated Src family member (predominantly Lyn (NISHIZUMI and YAMAMOTO 1997)), most likely via CD45 (BERGER et al. 1994), to tyrosine phosphorylate the β- and γ-ITAMs. SHIP is then attracted via its SH2 domain (DAMEN et al. 1996) to the C-terminal ITAM of the β- (KIMURA et al. 1997) and/or the γ (OSBORNE et al. 1996)-chain, becomes phosphorylated at one or both of its NPXY motifs (DAMEN et al. 1996) and then attracts Shc, via the latters PTB domain, to be phosphorylated. Inhibition of extracellular calcium entry by wortmannin or LY294002 suggests PI-3-kinase is

Fig. 4. Model of IgE-induced degranulation in SHIP$+/+$ and $-/-$ BMMCs. Although SHIP is shown associating with the ITAM of the β-subunit of the IgE receptor, it remains to be determined how SHIP is attracted to the plasma membrane following IgE stimulation. The ITAM's are depicted as *light gray bars* in the β- and γ-subunits. The SH2 domains of SHIP, Shc, Lyn, and Syk are shown in *black*

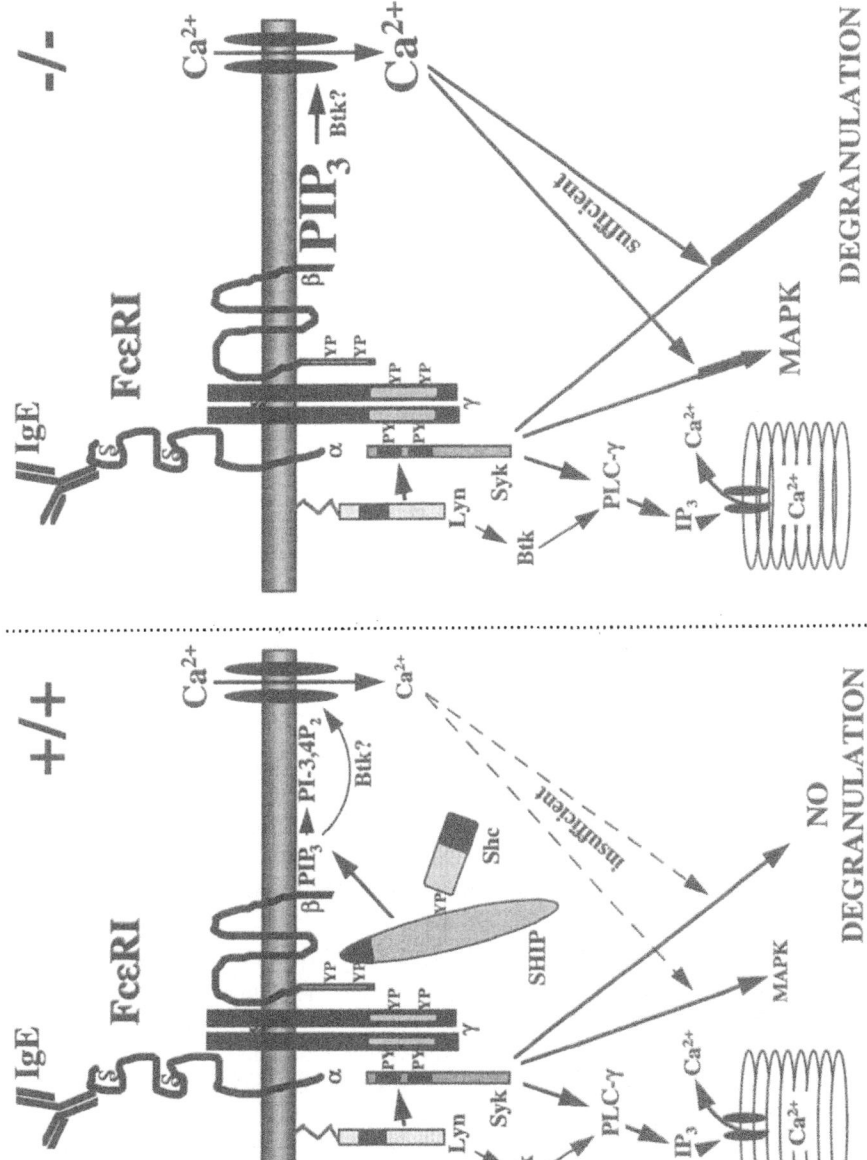

also translocated to the plasma membrane and activated early in this process but exactly how remains to be determined. Syk is attracted to the γ-ITAM (HIRASAWA et al. 1995), becomes phosphorylated by Lyn and stimulates the tyrosine phosphorylation of PLC-γ1 and γ2 (ZHANG et al. 1996), which in turn leads to the initial release of intracellular calcium via IP$_3$. This emptying of intracellular calcium triggers the entry of extracellular calcium and this entry is substantially augmented by PI-3-kinase mediated generation of PIP$_3$. As to how an elevation in PIP$_3$ can impact on extracellular calcium entry, two recent reports involving overexpression of various cDNAs in B cell lines suggest that activating the B cell receptor on B cells attracts the Btk/Tec family tyrosine kinase, Btk, (which has been shown to be activated in response to FcεR1 crosslinking in BMMCs (KAWAKAMI et al. 1994)) to PIP$_3$ at the plasma membrane. This then leads to the activation of Btk, its phosphorylation/activation of PLC-γ2 to generate IP$_3$ and the sustained emptying of intracellular calcium stores that are critical for activating the store operated calcium (SOC) entry from the extracellular medium (SCHARENBERG et al. 1998; FLUCKIGER et al. 1998). However, since we observe no detectable difference in the release of intracellular calcium in IgE-stimulated SHIP+/+ and −/− BMMCs (ie, in the presence of EGTA), we hypothesize that the elevated PIP$_3$ present in SHIP−/− cells attracts and activates Btk or another PH-containing intermediate at a step between the draining of intracellular calcium stores and extracellular calicum entry. This is in agreement with BOLLAND et al. (1998) who very recently concluded, from studying SHIP's role in mediating FcγRIIB inhibition of B cell receptor activation in the chicken DT40 B cell line, that SHIP regulates extracellular calcium entry after draining of intracellular calcium stores via its ability to control Btk association with the plasma membrane.

5 Concluding Remarks

Our results suggest that one of SHIP's roles is to both prevent inappropriate and limit appropriate IgE-mediated degranulation of mast cells by hydrolyzing PIP$_3$ and thus restricting the entry of extracellular calcium. This raises the intriguing possibility that naturally occurring mutations in SHIP could contribute to specific hyper-allergic conditions in man. During these studies we found with both SHIP+/+ and −/− mast cells, that IgE alone (ie, in the absence of cross-linkers) triggered both the entry of extracellular calcium and protein tyrosine phosphorylations. Although this may be due to low levels of IgE aggregates it suggests that the current paradigm of passive sensitization with IgE should be re-examined. It is conceivable, for example, that monomeric IgE triggers signaling by causing a conformational change in the multi-spanning β-subunit (in a fashion analogous to that proposed for seven spanner receptors). Intriguingly, this signaling by IgE alone may also occur in vivo, especially under conditions where serum IgE levels are markedly elevated, such as during allergic attacks or following specific infections, and serve to prime the cells for subsequent activation by multivalent antigens. Related to this, several reports have shown recently that IgE alone, both in vivo and in vitro, dramatically increases FcεR1 expression on the surface of mast cells and basophils (Hsu et al. 1996; YAMAGUCHI et al. 1997; MACGLASHAN Jr. et al. 1998). Moreover, FcεR1 expression on peritoneal mast cells from IgE−/− mice was dramatically reduced (greater than 80%) compared with that on cells from the corresponding normal mice (YAMAGUCHI et al. 1997). These results suggest that endogenous circulating IgE (whether it contains low levels of aggregates or not) is capable of initiating signaling. Thus, we would like to propose that IgE levels (up to 30µg/ml) that are reached in vivo following infections or allergic attacks may trigger signaling events in normal mast cells in the absence of multivalent antigens that actively prime the cells for subsequent degranulation by the antigen. Importantly, the presence of SHIP in these normal mast cells ensures that this priming step does not lead to degranulation.

Acknowledgments. We would like to thank Christine Kelly for typing the manuscript. This work was supported by the NCI-C and the MRC-C with core support from the British Columbia Cancer Foundation and the British Columbia Cancer Agency. MH is supported by the Deutsche Forschungsgemeinschaft. MPS holds a Cancer Research Society Studentship and VD is an MRC/BCLA Scientist. GK is a Terry Fox Cancer Research Scientist of the NCI-C supported by funds from the Canadian Cancer Society and the Terry Fox Run.

References

Alber G, Miller L, Jelsema CL, Varin-Blank N, Metzger H (1991) Structure-function relationships in the mast cell high affinity receptor for IgE. Role of the cytoplasmic domains and the beta subunit. J Biol Chem 266:22613–22620

Berger SA, Mak TW, Paige CJ (1994) The leukocyte common antigen (CD45) is required for immunoglobulin E-mediated degranulation of mast cells. J Exp Med 180:471–476

Bolland S, Pearse RN, Kurosaki T, Ravetch JV (1998) SHIP modulates immune receptor responses by regulating membrane association of Btk. Immunity 8:509–516

Chacko GW, Tridandapani S, Damen JE, Liu L, Krystal G, Coggeshall KM (1996) Negative signaling in B lymphocytes induces tyrosine phosphorylation of the 145-kDa inositol polyphosphate 5-phosphatase SHIP. J Immunol 157:2234–2238

Damen JE, Liu L, Rosten P, Humphries RK, Jefferson AB, Majerus PW, Krystal G (1996) The 145-kDa protein induced to associate with Shc by multiple cytokines is an inositol tetraphosphate and phosphatidylinositol 3,4,5-trisphosphate 5-phosphatase. Proc Natl Acad Sci USA 93:1689–1693

Damen JE, Liu L, Ware MD, Ermolaeva M, Majerus PW, Krystal G (1998) Multiple forms of the SH2-containing inositol phosphatase SHIP are generated by C-terminal truncation. Blood 92:1199–1205

Fluckiger A-C, Li Z, Kato RM, Wahl MI, Ochs HD, Longnecker R, Kinet J-P, Witte ON, Scharenberg AM, Rawlings DJ (1998) Btk/Tec kinases regulate sustained increases in intracellular Ca^{2+} following B-cell receptor activation. EMBO J 17:1973–1985

Geier SJ, Algate PA, Carlberg K, Flowers D, Friedman C, Trask B, Rohrschneider LR (1997) The human SHIP gene is differentially expressed in cell lineages of the bone marrow and blood. Blood 89:1876–1885

Helgason CD, Damen JE, Rosten P, Grewal R, Sorensen P, Chappel SM, Borowski A, Jirik F, Krystal G, Humphries RK (1998) Targeted disruption of SHIP leads to hemopoietic perturbations, lung pathology and a shortened lifespan. Genes & Dev 12:1610–1620

Hirasawa N, Scharenberg A, Yamamura H, Beaven MA, Kinet J-P (1995) A requirement for Syk in the activation of the microtubule-associated protein kinase/phospholipase A2 pathway by Fc-epsilon RI is not shared by a G protein-coupled receptor. J Biol Chem 270:10960–10967

Hsu C, MacGlashan D Jr (1996) IgE antibody up-regulates high affinity IgE binding on murine bone marrow-derived mast cells. Immunol Lett 52:129–134

Huber M, Helgason CD, Damen JE, Liu L, Humphries RK, Krystal G (1998) The src homology 2-containing inositol phosphatase (SHIP) is the gatekeeper of mast cell degranulation. Proc Natl Acad Sci USA (in press)

Irvine R (1998) Translocation translocation translocation. Curr Biol 8:R557-R559.

Johnson GR, Gonda TJ, Metcalf D, Hariharan IK, Cory S (1989) A lethal myeloproliferative syndrome in mice transplanted with bone marrow cells infected with a retrovirus expressing granulocyte-macrophage colony stimulating factor. EMBO J 8:441–448

Kavanaugh WM, Pot DA, Chin SM, Deuter-Reinhard M, Jefferson AB, Norris FA, Masiarz FR, Cousens LS, Majerus PW, Williams LT (1996) Multiple forms of an inositol polyphosphate 5-phosphatase form signaling complexes with Shc and Grb2. Curr Biol 6:438–445

Kawakami Y, Yao L, Miura T, Tsukada S, Witte ON, Kawakami T (1994) Tyrosine phosphorylation and activation of Bruton tyrosine kinase upon FcεRI cross-linking. Mol Cell Biol 14: 5108–5113

Kimura T, Sakamoto H, Appella E, Siraganian RP (1997) The negative signaling molecule SH2 domain-containing inositol-polyphosphate 5-phosphatase (SHIP) binds to the tyrosine-phosphorylated beta subunit of the high affinity IgE receptor. J Biol Chem 272:13991–13996

Kuroiwa A, Yamashita Y, Inui M, Yuasa T, Ono M, Nagabukuro A, Matsuda Y, Takai T (1998) Association of tyrosine phosphatases SHP-1 and SHP-2 inositol 5-phosphatase SHIP with gp49B1 and chromosomal assignment of the gene. J Biol Chem 273:1070–1074

Lamkin TD, Walk SF, Liu L, Damen JE, Krystal G, Ravichandran KS (1997) Shc interaction with Src homology 2 domain containing inositol phosphatase (SHIP) in vivo requires the Shc-phosphotyrosine binding domain and two specific phosphotyrosines on SHIP. J Biol Chem 272:10396–10401

Lioubin MN, Algate PA, Tsai S, Carlberg K, Aebersold A, Rohrschneider LR (1996) p150^{Ship} a signal transduction molecule with inositol polyphosphate-5-phosphatase activity. Genes & Develop 10:1084–1095

Liu L, Damen JE, Ware M, Hughes M, Krystal G (1997) SHIP a new player in cytokine-induced signaling. Leukemia 11:181–184

Liu L, Damen JE, Hughes MR, Babic I, Jirik FR, Krystal G (1997a) The Src homology 2 (SH2) domain of SH2-containing inositol phosphatase (SHIP) is essential for tyrosine phosphorylation of SHIP its association with Shc and its induction of apoptosis. J Biol Chem 272:8983–8988

Liu L, Damen JE, Ware MD, Krystal G (1997b) Interleukin-3 induces the association of the inositol 5-phosphatase SHIP with SHP2. J Biol Chem 272:10998–11001

Liu Q, Shalaby F, Jones J, Bouchard D, Dumont DJ (1998) The SH2-containing inositol polyphosphate 5-phosphatase Ship is expressed during hematopoiesis and spermatogenesis. Blood 91:2753–2759

MacGlashan D Jr, McKenzie-White J, Chichester K, Bochner BS, Davis FM, Schroeder JT, Lichtenstein LM (1998) In vitro regulation of FcεRIα expression on human basophils by IgE antibody. Blood 91:1633–1643

Nishizumi H, Yamamoto T (1997) Impaired tyrosine phosphorylation and Ca^{2+} mobilization but not degranulation in lyn-deficient bone-marrow-derived mast cells. J Immunol 158:2350–2355

Odai H, Sasaki K, Iwamatsu A, Nakamoto T, Ueno H, Yamagata T, Mitani K, Yazaki Y, Hirai H (1997) Purification and molecular cloning of SH2- and SH3-containing inositol polyphosphate-5-phosphatase which is involved in the signaling pathway of granulocyte-macrophage colony-stimulating factor erythropoietin and Bcr-Abl. Blood 89:2745–2756

Ono M, Bolland S, Tempst P, Ravetch JV (1996) Role of the inositol phosphatase SHIP in negative regulation of the immune system by the receptor Fc-gamma-RIIB. Nature 383:263–266

Ono M, Okada H, Bolland S, Yanagi S, Kurosaki T, Ravetch JV (1997) Deletion of SHIP or SHP-1 reveals two distinct pathways for inhibitory signaling. Cell 90:293–301

Osborne MA, Zenner G, Lubinus M, Zhang X, Songyang Z, Cantley LC, Majerus P, Burn P, Kochan JP (1996) The inositol 5′-phosphatase SHIP binds to immunoreceptor signaling motifs and responds to high affinity IgE receptor aggregation. J Biol Chem 271:29271–29278

Sattler M, Salgia R, Shrikhande G, Verma S, Choi J-L, Rohrschneider LR, Griffin JD (1997) The phosphatidylinositol polyphosphate 5-phosphatase SHIP and the protein tyrosine phosphatase SHP-2 form a complex in hematopoietic cells which can be regulated by BCR/ABL and growth factors. Oncogene 15:2379–2384

Scharenberg AM, El-Hillal O, Fruman DA, Beitz LO, Li Z, Lin S, Gout I, Cantley LC, Rawlings DJ, Kinet J-P (1998) Phosphatidylinositol-3,4,5-trisphosphate (PtdIns-3,4,5-P_3)/Tec kinase-dependent calcium signaling pathway a target for SHIP-mediated inhibitory signals. EMBO J 17:1961–1972

Scheid MP, Duronio V (1996) Phosphatidylinositol 3-OH kinase activity is not required for the activation of mitogen-activated protein kinase by cytokines. J Biol Chem 271:18134–18139

Tridandapani S, Kelley T, Pradhan M, Cooney D, Justement LB, Coggeshall KM (1997) Recruitment and phosphorylation of SH2-containing inositol phosphatase and Shc to the B-cell Fc gamma immunoreceptor tyrosine-based inhibition motif peptide motif. Mol Cell Biol 17:4305–4311

Vély F, Olivero S, Olcese L, Moretta A, Damen JE, Liu L, Krystal G, Cambier JC, Daëron M, Vivier E (1997) Differential association of phosphatases with hematopoietic co-receptors bearing immunoreceptor tyrosine-based inhibition motifs. Eur J Immunol 27:1994–2000

Ware MD, Rosten P, Damen JE, Liu L, Humphries RK, Krystal G, (1996) Cloning and characterization of human SHIP the 145-kD inositol 5-phosphatase that associates with SHC after cytokine stimulation. Blood 88:2833–2840

Yamaguchi M, Lantz CS, Oettgen HC, Katona IM, Fleming T, Miyajima I, Kinet J-P, Galli SJ (1997) IgE enhances mouse mast cell Fc(epsilon)RI expression in vitro and in vivo: evidence for a novel amplification mechanism in IgE-dependent reactions. J Exp Med 185:663–672

Yano H, Nakanishi S, Kimura K, Hanai N, Saitoh Y, Fukui Y, Nonomura Y, Matsuda Y (1993) Inhibition of histamine secretion by wortmannin through the blockade of phosphatidylinositol 3-kinase in RBL-2H3 cells. J Biol Chem 268:25846–25856

Zhang J, Berenstein EH, Evans RL, Siraganian RP (1996) Transfection of Syk protein tyrosine kinase reconstitutes high affinity IgE receptor-mediated degranulation in a Syk-negative variant of rat basophilic leukemia RBL-2H3 cells. J Exp Med 184:71–79

The Unexpected Complexity of FcγRIIB Signal Transduction

J.C. Cambier, D. Fong, and I. Tamir

1 Inhibitory Receptors: Concept, Function, and Distribution

Recent studies have defined a family of modulatory receptors and coreceptors that function to inhibit the transduction of certain activating signals. These receptors, which now number more than 75, are structurally·similar and employ similar signaling effectors, but are found in diverse cellular systems (for review see CAMBIER 1997; VELY and VIVIER 1997; LANIER 1998; YOKOYAMA 1998). The longest-recognized and most extensively studied member is FcγRIIB, a low-affinity receptor for immunoglobulin G (IgG) constant regions that occurs in three alternatively spliced isoforms which differ only in their cytoplasmic tail sequence (see Fig. 1) (FRIDMAN, BONNEROT et al. 1992; LATOUR, FRIDMAN et al. 1996). All isoforms mediate transduction of inhibitory signals. The FcγRIIB1 isoform is expressed by B lymphocytes throughout the lineage, as well as by certain mast cells, other granulocytes, and activated T cells. When coaggregated with antigen receptors, e.g. BCR and FcεRI, by immune complexes containing IgG antibodies, FcγRIIB1 mediate inhibition of antigen receptor-mediated B cell blastogenesis and proliferation, and divert the response to apoptosis (CHAN and SINCLAIR 1971; PHILLIPS and PARKER 1983; PHILLIPS and PARKER 1984; ASHMAN, PECKHAM et al. 1996; ONO, OKADA et al. 1997). Other prominent members of the

Division of Basic Sciences, Department of Pediatrics, National Jewish Center for Immunology and Respiratory Medicine, 1400 Jackson Street, Denver, CO 80206, USA; and Department of Immunology, University of Colorado Health Sciences Center, Denver, CO 80206, USA

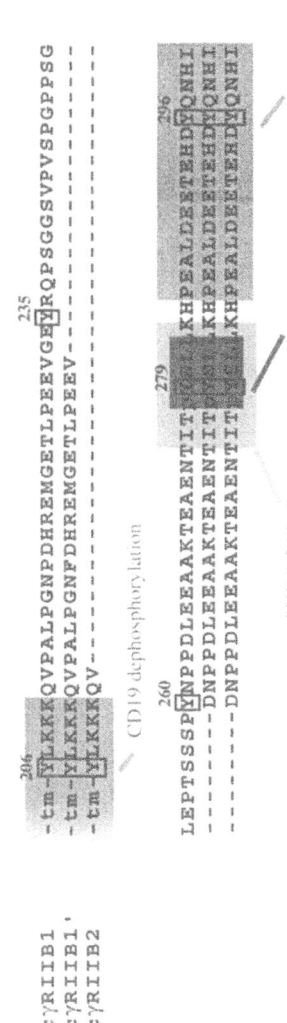

Fig. 1. Sequence comparison of murine FcγRIIB1, FcγRIIB1′ and FcγRIIB2 cytoplasmic tails with annotation of functional sites

inhibitory receptor family are the killer inhibitory receptors (KIR) that function to block target cell killing by natural killer (NK) cells, certain T cells (LANIER 1998; YOKOYAMA 1998), and CTLA4, that functions late in the T cell response to attenuate clonal outgrowth (THOMPSON and ALLISON 1997). Signal inhibitory regulator proteins (SIRPs), function outside the immune system, inhibiting signals transduced by PDGF, EGF, and insulin receptors (KHARITONENKOV, CHEN et al. 1997).

As noted above, members of this inhibitory receptor family are structurally similar. All are single membrane spanners, with C-type lectin or immunoglobulin-like extracellular domains. Their cytoplasmic domains contain one or more copies of the immunoreceptor tyrosine-based inhibition motif (ITIM), the consensus of which is I/VxYxxL. This motif mediates receptor associations with the protein tyrosine phosphatases SHP-1 and/or SHP-2 in all members of the family (CAMBIER 1997). However, for a small subset that includes FcγRIIB isoforms, p91PIR-B, and gp49B1, the ITIM also mediates receptor association with the SH2 domain-containing inositol 5′-phosphatase SHIP (D'AMBROSIO, FONG et al. 1996; DAMEN, LIU et al. 1996; KUROIWA, YAMASHITA et al. 1998; YAMASHITA, FUKUTA et al. 1998). ITIMs are also found in the cytoplasmic tails of certain cytokine receptors where they may function in feedback regulation.

The physiology function of FcγRIIB remains only partially defined. A number of laboratories showed in the late 1960s that under certain circumstances immune complexes containing IgG antibodies are not immunogenic in animals and that this lack of immunogenicity is not due to antibody blocking of immunogenic epitopes (CHAN and SINCLAIR 1971; FRIDMAN, BONNEROT et al. 1992). It was shown in the 1970s that antigens containing immunoglobulin G Fc regions were very potent B cell tolerogens (WALDSCHMIDT and VITETTA 1985). Coaggregation of FcγRIIB with BCR leads to inhibition of BCR-mediated blastogenesis and proliferation (PHILLIPS and PARKER 1983), and to apoptosis (ASHMAN, PECKHAM et al. 1996). Finally, FcγRIIB1 coaggregation with BCR inhibits B cell CD69 and CD86 expression (Benschop and Cambier, unpublished observations). This inhibitory Fc receptor function was shown to be defective in autoimmunity-prone NZB mice, raising the possibility that it plays a role in preventing expansion and/or differentiation of autoreactive B cells (UHER and DICKLER 1986). In 1996, an FcγRIIB knockout mouse was reported by Takai and colleagues (TAKAI, ONO et al. 1996). It exhibits five- to tenfold enhanced antibody production, suggestive of impaired regulation of the immune response; and enhanced sensitivity to IgG-mediated cutaneous ana-phylaxis, consistent with absence of inhibitory Fcγ receptor function in mast cells. Taken together, these findings suggest that inhibitory Fc receptors play an important role in dampening the antibody response once IgG antibodies are produced, lessening the likelihood of autoantibody production. They also raise the possibility that agonists of intermediaries in inhibitory signaling pathways might be of therapeutic value in autoimmunity and transplantation while antagonists might be useful adjuvants.

2 Molecular Mechanisms of FcγRIIB Signaling

Initial insight regarding the molecular bases of inhibitory Fc receptor function came with observations that FcγRIIB1 can exert inhibitory effects at least in part at the level of early events in BCR signaling. Wilson et al. (WILSON, GREENBLATT et al. 1987) showed in 1987 that FcγRIIB coaggregation leads to inhibition of BCR-mediated extracellular calcium influx. Bijsterbosch and Klaus (BIJSTERBOSCH and KLAUS 1985) showed that coaggregation of these receptors leads to inhibition of BCR-mediated inositol trisphosphate (Ins1,4,5P₃) production. These findings indicated that one site of FcγRIIB1 effector action is at or proximal to BCR activation of phospholipase C gamma (PLCγ). In addition, it has been shown that coaggregation of these receptors leads to inhibition of BCR-mediated $p21^{ras}$ activation (TRIDANDAPANI, CHACKO et al. 1997) and increased tyrosyl phosphorylation of SHIP (CHACKO, TRIDANDAPANI et al. 1996; TRIDANDAPANI, KELLEY et al. 1997). The role of these early molecular events in later FcγRIIB-mediated inhibition of blastogenesis and proliferation, and activation of apoptosis is unknown.

Impetus for our initial approach to elucidation of the molecular basis of FcγRIIB action came from observations by Amigorena et al. (AMIGORENA, BON-NEROT et al. 1992) that a specific FcγRIIB1 cytoplasmic tail sequence is necessary for its signaling function, and from findings by Muta and Ravetch (MUTA, KUR-OSAKI et al. 1994) that this sequence alone could mediate inhibitory signaling when placed in an inert receptor context. Pursuant to earlier success in defining Igα and Igβ effectors based on their binding to phosphopeptides derived from the B cell antigen receptor (CLARK, CAMPBELL et al. 1992), we assessed the binding of cellular proteins to tyrosyl-phosphorylated FγRIIB1 peptides synthesized based on findings of Amigorena (AMIGORENA, BONNEROT et al. 1992) and Muta (MUTA, KUROSAKI et al. 1994). Surprisingly, significant binding of only three proteins, SHP-1, SHP-2, and SHIP, was detected based on S^{35}-methionine labeling (D'AMBROSIO, HIPPEN et al. 1995; D'AMBROSIO, FONG et al. 1996). Measurement of their in vivo association with FcγRIIB1 by coimmunoprecipitation suggested that the predominant associating protein is SHIP (ONO, BOLLAND et al. 1996; Nakamura and Cambier, unpublished observations). Nonetheless, analysis of FcγRIIB function in motheaten mice lacking functional SHP-1 revealed that this tyrosine phosphatase is necessary for FcγRIIB inhibition of BCR-mediated B cell proliferation (D'AM-BROSIO, HIPPEN et al. 1995). However, SHP-1 was not necessary for receptor-mediated inhibition of Ca^{2+} mobilization (NADLER, CHEN et al. 1997).

Relying on findings implicating protein phosphatases in FcγRIIB1 function, we embarked on analysis of receptor modulation of the early tyrosine phosphorylation events that follow BCR aggregation and effect PLCγ activation (Fig. 2). To our surprise, phosphorylation of the known intermediaries Lyn, Igα, Igβ, and Syk was not affected, at least at time points corresponding to the calcium mobilization response (BUHL, PLEIMAN et al. 1997; HIPPEN, BUHL et al. 1997). In most experiments, we saw no effect of FcγRIIB coaggregation on BCR-mediated PLCγ phosphorylation. However, effects have been reported by SARKAR, SCHLOTTMANN

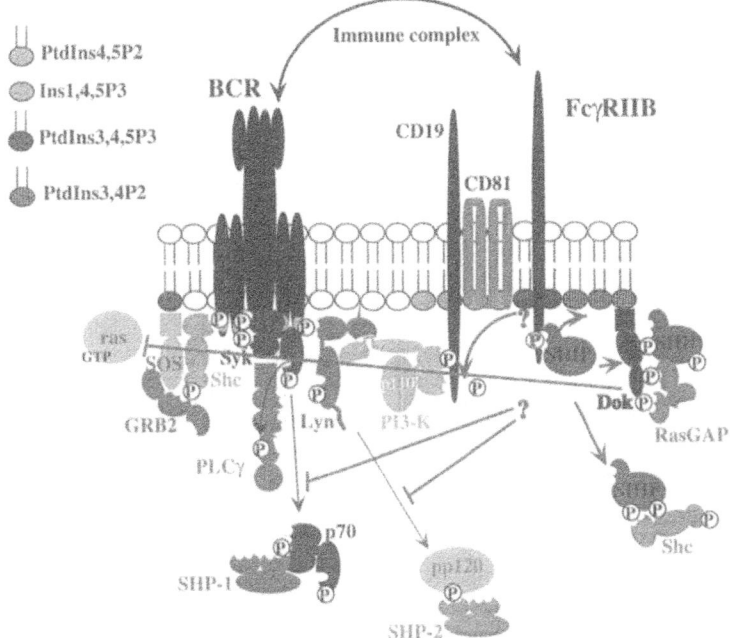

Fig. 2. Schematic of BCR signal transduction and inhibition by FcγRIIB

et al. (1996). These data suggest that FcγRIIB affects some process that is needed, perhaps in addition to PLCγ phosphorylation, for PLCγ-mediated hydrolysis of inositol phospholipids.

When the tyrosyl phosphorylation of total cellular proteins was analyzed by immunoblotting of whole cell lysates, we noted an 80% reduction in phosphorylation of one of the major substrates in 15s, with complete loss of phosphorylation within 45s following FcγRIIB-BCR coaggregation, as compared to BCR aggregation alone (HIPPEN, BUHL et al. 1997). We identified this protein as CD19, a B cell-restricted cell surface glycoprotein that is expressed throughout the lineage (TEDDER, ZHOU et al. 1994; FEARON and CARTER 1995). Similar observations were made by Kiener et al. (KIENER, LIOUBIN et al. 1997). CD19 is expressed on the cell surface in two forms; while most is expressed in association with CD81, a minority of about 20% is expressed in association with both CD81 and CD21, constituting the type 2 complement receptor complex CR2. CD21 mediates C3dg binding, while CD81, a tetraspan molecule, and CD19 have no known ligands. CD19 reportedly associates with the BCR and is a single membrane-spanning molecule with two full and one partial Ig-like domains in its extracellular aspect (PESANDO, BOUCHARD et al. 1989; CARTER, DOODY et al. 1997). Its cytoplasmic tail contains nine tyrosines that are potential phosphorylation sites, which may act to recruit Lyn, Vav, and phosphatidylinositol 3 kinase (PI3-k) following BCR-CD19 coligation (TUVESON, CARTER et al. 1993; van NOESEL, LANKESTER et al. 1993; WENG, JARVIS et al. 1994;

O'ROURKE, TOOZE et al. 1998). Our studies indicate that only two CD19 tyrosines, Y484 and Y515, are phosphorylated upon aggregation of BCR alone (Buhl and Cambier, submitted for publication). These tyrosines have been shown to mediate CD19 association with phosphatidylinositol 3-kinase (TUVESON, CARTER et al. 1993). Consistent with the ability of FcγRIIB coaggregation with BCR to prevent or reverse the phosphorylation of these tyrosines, we showed that this coaggregation leads to failed recruitment of PI3-k and failed BCR-mediated PI3-k activation (HIPPEN, BUHL et al. 1997). Together with findings of Bijsterbosch and Klaus (BIJSTERBOSCH, KLAUS 1985) and Wilson et al. (WILSON, GREENBLATT et al. 1987), results suggested that phosphatidylinositol 3,4,5 triphosphate (PtdIns3,4,5P$_3$), the product of PtdIns4,5P$_2$ phosphorylation by PI3-k, is required for BCR-mediated phosphoinositide hydrolysis by PLCγ. This was confirmed by studies demonstrating that BCR-mediated Ins1,4,5P$_3$ production and calcium mobilization are both inhibited by the PI3-k inhibitors wortmannin and LY294002, and are dependent on cellular expression of CD19 (BUHL, PLEIMAN et al. 1997; HIPPEN, BUHL et al. 1997). Furthermore, CD19 support of BCR- mediated PI3-k activation, Ins1,4,5P$_3$ production, and calcium mobilization responses required that CD19 contain Y484 and Y515, the PI3-k "integrators" (BUHL, PLEIMAN et al. 1997 and Buhl and Cambier, submitted for publication).

In recent months it has been reported that hydrolysis of PtdIns4,5P$_2$ requires that PLCγ be translocated to the plasma membrane via binding of its pleckstrin homology (PH) domain to PtdIns3,4,5P$_3$ (FALASCA, LOGAN et al. 1998). Recognition of this requirement appeared to complete circuitry between FcγRIIB1-mediated CD19 dephosphorylation and FcγRIIB1 inhibition of phosphoinositide hydrolysis. However, it appears that at least one additional intermediary exists. It has been known for some time that the pleckstrin homology domain of Bruton's tyrosine kinase (Btk) binds avidly to PtdIns3,4,5P$_3$ (SALIM, BOTTOMLEY et al. 1996) and that Btk is required for BCR activation of Ins1,4,5P$_3$ production and calcium mobilization (RIGLEY, HARNETT et al. 1989; TAKATA, KUROSAKI 1996). These findings suggested that Btk may somehow cooperate with PLCγ during BCR signal transduction, and that this cooperativity is dependent on PtdIns3,4,5P$_3$ production and, thus, CD19 phosphorylation. In recent studies we have found that BCR-mediated Btk activation is dependent on CD19 expression and inhibited by wortmannin (Buhl and Cambier, submitted for publication). Importantly, Bolland et al. (BOLLAND, PEARSE et al. 1998) have shown that expression of membrane-targeted Btk can, at least partially, rescue FcγRIIB1 inhibition of BCR-mediated calcium mobilization.

Taken together, these data support the model shown in Fig. 2, which posits that BCR-mediated signaling proceeds via the aggregation-driven activation of Igα/Igβ-associated Src-family kinases, predominantly Lyn. These kinases phosphorylate the receptor's immunoreceptor tyrosine-based activation motifs (ITAMs), leading to recruitment of Syk, additional Src family kinases, Shc (D'AMBROSIO, HIPPEN et al. 1996), and presumably other SH2-containing effectors (CAMBIER 1995). Apposition of these molecules promotes transphosphorylation, leading to further recruitment of effectors such as: PI3-k, that binds via its proline-rich regions

(PLEIMAN, HERTZ et al. 1994) to the Lyn SH3 domain; Grb2-SOS, that associates via Grb2 SH2 domains with Shc phosphotyrosyl residues; and PLCγ, that associates via its SH2 domains with phospho-tyrosyl residues on Syk (SIDORENKO, LAW et al. 1995; SILLMAN and MONROE 1995). These interactions facilitate activation of Src- and Syk-family kinases and the $p21^{ras}$ (ras) pathway, and limited (~10% of normal) activation of PI3-k. CD19 associated with the BCR (PESANDO, BOUCHARD et al. 1989) is phosphorylated by BCR-activated tyrosine kinases. The tyrosyl phosphorylated CD19 binds to the p85 subunit of PI3-k via the latter's SH2 domains (possibly those within PI3-k already associated with Lyn), activating PI3-k by a steric effect on p110 catalytic subunit (TUVESON, CARTER et al. 1993). The consequent generation of $PtdIns3,4,5P_3$ leads to recruitment of PLCγ and Btk to the plasma membrane. Subsequent Btk activation by Lyn-mediated phosphorylation and autophosphorylation leads to PLCγ phosphorylation, further promoting its hydrolysis of phosphoinositides (LI, RAWLINGS et al. 1997; LI, WAHL et al. 1997; SATTERTHWAITE, CHEROUTRE et al. 1997; WAHL, FLUCKIGER et al. 1997; FLUCKIGER, LI et al. 1998; SCHARENBERG, EL-HILLAL et al. 1998).

As also shown in Fig. 2, coaggregation of FcγRIIB with BCR leads to ITIM tyrosine phosphorylation and activation of multiple effectors, including those that mediate dephosphorylation or prevent phosphorylation of CD19. This inhibits PI3-k activation, blocking $PtdIns3,4,5P_3$-dependent downstream processes such as Btk and PLCγ activation. Considerable evidence indicates that recruitment of SHIP to phosphorylated FcγRIIB also functions to inhibit Btk and PLCγ activation, in this case by promoting $PtdIns3,4,5P_3$ hydrolysis (BOLLAND, PEARSE et al. 1998; SCHARENBERG, EL-HILLAL et al. 1998). Recent studies indicate that during BCR signaling, SHP-1 becomes associated with unidentified tyrosyl-phosphorylated 70-kDa and 41-kDa proteins, while SHP-2 becomes associated with an unidentified 120kDa phosphoprotein (NAKAMURA and CAMBIER 1998; Nakamura and Cambier, unpublished observations). Under conditions of FcγRIIB-BCR coaggregation, either these interactions are inhibited or the associated molecules are dephosphorylated (NAKAMURA and CAMBIER 1998).

3 Linker Function of SHIP During FcγRIIB Signaling

The function of SHIP in inhibitory signal transduction may depend not only on its enzymatic activity but also on its function as a linker molecule. SHIP becomes tyrosyl-phosphorylated upon binding to FcγRIIB. Subsequent dissociation is related to its binding to Shc (PRADHAN and COGGESHALL 1997; TRIDANDAPANI, KELLEY et al. 1997) and p62Dok (Dok), and phosphorylation of these molecules (Tamir, Nakamura, and Cambier, submitted for publication). Phosphorylation of Dok following FcγRIIB-BCR coaggregation is correlated with enhanced binding rasGAP. It has been suggested by Coggeshall and colleagues that Shc binding to the SHIP may inhibit ras activation by sequestering Shc needed by the BCR for ras activation. However, we believe it more likely that rasGAP recruitment to the coaggregated receptor complex inactivates ras by activation of its GTPase.

4 Mutational Analysis of FcγRIIB Signaling Function

Although the effect of FcγRIIB in reducing tyrosyl phosphorylation of CD19 suggests the involvement of a protein tyrosine phosphatase, no direct evidence exists that the observed effect is phosphatase-mediated. Alternatively, this effect could reflect cessation of CD19 phosphorylation due to inhibition of a kinase or disruption of BCR-CD19 association, following which constitutively active "housekeeping" phosphatases may dephosphorylate CD19.

We initiated two types of mutational analysis to define the molecular basis of the effect of FcγRIIB1 on CD19 phosphorylation and its functional consequence relative to other FcγRIIB effectors such as SHIP. The goal was to create FcγRIIB1 receptors that would engage different subsets of effectors and use these to study downstream functions of the respective effectors. Firstly, in collaboration with Vivier and colleagues (VELY, OLIVERO et al. 1997), we explored ITIM sequence requirements for interaction with SHP-1, SHP-2, and SHIP. As also shown by Burshtyn et al. (BURSHTYN, YANG et al. 1997), these studies revealed that the hydrophobic residue at Y-2 position (I277) (Fig. 1) is critical for SHP-1 and SHP-2 binding but not for that of SHIP. This provided a potential approach for generating receptors that can engage SHIP but not SHP-1 or SHP-2.

We then initiated formal mutational analysis of FcγRIIB tail function (Fong, Minskoff et al. submitted for publication). FcγRIIB1, B1′ and B2 splice variants, and various mutants were expressed in IIA1.6, an FcγR-negative variant of the A20 B lymphoma (JONES, TITE et al. 1986). Analysis of mutants revealed the surprising result that the ability of FcγRIIB1 to mediate dephosphorylation of CD19 is completely independent of the ITIM. Based on the ability of B1 truncation mutants (CT259, a cytoplasmic domain truncation to residue 259) as well as B1′ and B2 isoforms to mediate this effect (see Fig. 1), requisite information must be contained in the extracellular and/or transmembrane domains (which are identical in B1, B1′ and B2), or the six juxtamembrane residues in the cytoplasmic tail. Consistent with the above-described CD19-PLCγ circuitry, these regions of the molecule contain information needed to inhibit PtdIns3,4,5P_3 production by 40% and calcium mobilization responses by approximately 50%. Using the I277 A ITIM mutant, loss of the SHP-1 binding function was found to have no significant effect on the ability of FcγRIIB to inhibit calcium mobilization. These results are consistent with findings of Nadler et al. (NADLER, CHEN et al. 1997) that FcγRIIB1 inhibition of Ca^{2+} mobilization and CD19 dephosphorylation is seen in B cell lines derived from SHP-1 negative motheaten mice and findings of Gupta et al. and Ono et al. that SHP-1 is required for KIR but not FcγRIIB inhibition of Ca^{2+} mobilization (GUPTA, SCHARENBERG et al. 1997; ONO, OKADA et al. 1997). It is noteworthy that FcγRIIB chimeras, in which the receptor tail is replaced with either the SHP-1 or SHIP catalytic domains, can inhibit Ca^{2+} mobilization, implicating both SHP-1 and SHIP as mediators of ITIM effects on the calcium response (ONO, OKADA et al. 1997). Thus three distinct FcγRIIB-activated pathways can modulate calcium mobilization.

Analysis of the ITIM mutants Y279 A and LL281/282AA that cannot engage SHIP, SHP-1, or SHP-2 revealed, by comparison with I277 A, that SHIP plays a significant role in reduction of $PtdIns3,4,5P_3$ to unstimulated cell levels and is required for complete inhibition of the calcium response seen when wild type is coaggregated with the BCR FcγRIIB1. This result is consistent with observations that most FcγRIIB1 inhibition of Ca^{2+} mobilization is lost in SHIP gene-ablated DT40 (chicken B lymphoma) cells (ONO, OKADA et al. 1997).

Our mutational analyses revealed two additional, unexpected findings. The first is that a site distal from the ITIM in the FcγRIIB tail is necessary for "stable" association of SHIP with FcγRIIB, detectable by coprecipitation. The second is that non-ITIM tyrosines 260 and 296 are also phosphorylated upon FcγRIIB1-BCR coaggregation.

5 Integration of SHP-1, SHP-2, SHIP and CD19-Dephosphorylation Effects

It is obvious that FcγRIIB signaling is much more complex than was first thought (Fig. 2). Inhibitory functions are encoded in multiple regions of the molecule, and at least three distinct effector mechanisms can target the BCR-mediated calcium response. It appears that both CD19-dephosphorylation and SHIP-involving mechanisms target $PtdIns3,4,5P_3$. CD19 dephosphorylation prevents $PtdIns3,4,5P_3$ generation while SHIP cleaves it, yielding $PtdIn3,4P_2$.

The mechanism by which SHP-1 mediates inhibition of Ca^{2+} mobilization is less clear. It seems likely, as previously suggested (BINSTADT, BRUMBAUGH et al. 1996; LEIBSON 1997), that SHP-1 affects phosphorylation of early intermediaries in antigen receptor signaling pathways. Dephosphorylation of receptor ITAM tyrosines and/or Syk-family tyrosine kinases, as was shown to occur upon KIR inhibition of NK activation, could obviously prevent receptor-mediated activation of multiple (all) downstream pathways. In view of this, it is somewhat surprising that FcγRIIB1 does not significantly affect BCR-mediated Igα, Igβ, and Syk phosphorylation. Interestingly, surface plasmon resonance analysis indicates that while SHIP binds the phosphorylated FcγRIIB1 phosphopeptides with fast-on, fast-off kinetics, SHP-1 binds with slow-on, slow-off kinetics (Famiglietti and Cambier, unpublished observations). These findings suggest that SHP-1 may gradually displace SHIP following FcγRIIB phosphorylation and thus may function later in the response to dephosphorylate an intermediary, e.g. Syk and BCR, whose phosphorylation is required for the proliferative response. SHP-1 association with Syk could be involved in this inhibitory function.

The role of SHP-2 in inhibitory signaling is even more obscure. SHP-2 has been implicated in inhibitory signaling by CTLA4 (MARENGERE, WATERHOUSE et al. 1996; THOMPSON and ALLISON 1997), but recent studies by Frearson et al. using loss-of-function SHP-2 mutants indicate that it plays a positive role in T-cell antigen receptor signaling (D. Alexander, personal communication). As noted above, we observed that BCR aggregation stimulates SHP-2 association with an unidentified cytosolic phosphoprotein of about 120kDa (pp120), and this association/phosphorylation is inhibited by FcγRIIB1 coaggregation with BCR (NAKAMURA and CAMBIER 1998). Interestingly, TCR ligation leads to SHP-1 association with what may be the same 120 kDa phosphoprotein (FREARSON, YI et al. 1996). This interaction may play a positive role in antigen receptor signaling which is inhibited upon its disruption.

References

Amigorena S, Bonnerot C et al. (1992) Cytoplasmic domain heterogeneity and functions of IgG Fc receptors in B lymphocytes. Science 256(5065):1808–1812

Ashman RF, Peckham D et al. (1996). Fc receptor off-signal in the B cell involves apoptosis. Journal of Immunology 157:5–11

Bijsterbosch MK, Klaus GG (1985) Crosslinking of surface immunoglobulin and Fc receptors on B lymphocytes inhibits stimulation of inositol phospholipid breakdown via the antigen receptors. Journal of Experimental Medicine 162(6):1825–1836

Binstadt BA, Brumbaugh KM et al. (1996) Sequential involvement of Lck and SHP-1 with MHC-recognizing receptors on NK cells inhibits FcR-initiated tyrosine kinase activation. Immunity 5: 629–638

Bolland S, Pearse R et al. (1998) SHIP modulates immune receptor responses by regulating membrane association of Btk. Immunity 8(4):509–516.

Buhl AM, Pleiman CM et al. (1997) Qualitative regulation of B cell antigen receptor signaling by CD19: selective requirement for PI3-kinase activation, inositol-1,4,5-trisphosphate production, and Ca^{2+} mobilization. Journal of Experimental Medicine 186(11):1897–1910

Burshtyn DN, Yang W et al. (1997) A novel phosphotyrosine motif with a critical amino acid at position-2 for the SH2 domain-mediated activation of the tyrosine phosphatase SHP-1. Journal of Biological Chemistry 272:13066–13072

Cambier JC (1995) Antigen and Fc receptor signaling. The awesome power of the immunoreceptor tyrosine-based activation motif (ITAM). Journal of Immunology 155(7):3281–3285

Cambier JC (1997) Inhibitory receptors abound? Proceedings of the National Academy of Sciences, USA 94:5993–5995

Carter RH, Doody GM et al. (1997) Membrane IgM-Induced Tyrosine Phosphorylation of CD19 Requires a CD19 Domain That Mediates Association with Components of the B Cell Antigen Receptor Complex. Journal of Immunology 158:3062–3069

Chacko GW, Tridandapani S et al. (1996) Negative signaling in B lymphocytes induces tyrosine phosphorylation of the 145-kDa inositol polyphosphate 5-phosphatase, SHIP. J Immunol 157(6): 2234–8

Chan PL, Sinclair NRSC (1971) Regulation of the immune response: V. An analysis of the function of the Fc portion of antibody in suppression of an immune response with respect to interaction with components of the lymphoid system. Immunology 21:967–981

Clark MR, Campbell KS et al. (1992) The B cell antigen receptor complex: association of Ig-alpha and Ig-beta with distinct cytoplasmic effectors. Science 258(5079):123–6

D'Ambrosio D, Fong DC et al. (1996) The SHIP phosphatase becomes associated with Fc gamma RIIB1 and is tyrosine-phosphorylated during 'negative' signaling. Immunology Letters 54(2–3):77–82

D'Ambrosio D, Hippen KL et al. (1996) Distinct mechanisms mediate SHC association with the activated and resting B cell antigen receptor. European Journal of Immunology 26(8):1960–5

D'Ambrosio D, Hippen KL et al. (1995) Recruitment and activation of PTP1 C in negative regulation of antigen receptor signaling by Fc gamma RIIB1 [see comments]. Science 268(5208):293–7

Damen JE, Liu L et al. (1996) The 145-kDa protein induced to associate with Shc by multiple cytokines is an inositol tetraphosphate and phosphatidylinositol 3,4,5-triphosphate 5-phosphatase. Proceedings of the National Academy of Sciences of the United States of America 93(4):1689–93

Falasca M, Logan SK et al. (1998) Activation of phospholipase C gamma by PI 3-kinase-induced PH domain-mediated membrane targeting. EMBO Journal 17(2):414–22

Fearon DT, Carter RH (1995) The CD19/CR2/TAPA-1 complex of B lymphocytes: linking natural to acquired immunity. [Review]. Annual Review of Immunology 13(127):127–149.

Fluckiger A, Li Z et al. (1998) Btk/Tec kinases regulate sustained increases in intracellular Ca^{2+} following B-cell receptor activation. EMBO Journal 17(7):1973–1985

Frearson JA, Yi T et al. (1996) A tyrosine-phosphorylated 110–120-kDa protein associates with the C-terminal SH2 domain of phosphotyrosine phosphatase-1D in T cell receptor-stimulated T cells. European Journal of Immunology 26(7):1539–43

Fridman WH, Bonnerot C et al. (1992) Structural bases of Fc gamma receptor functions. Immunol Rev 125:49–76

Gupta N, Scharenberg AM et al. (1997) Negative signaling pathways of the killer cell inhibitory receptor and Fc gamma RIIb1 require distinct phosphatases. Journal of Experimental Medicine 186(3):473–8

Hippen KL, Buhl AM et al. (1997) FcγRIIB1 inhibition of BCR mediated phosphoinositide hydrolysis and Ca^{2+} mobilization is integrated by CD19 dephosphorylation. Immunity 7:49–58

Jones B, Tite JP et al. (1986) Different phenotypic variants of the mouse B cell tumor A20/2 J are selected by antigen- and mitogen-triggered cytotoxicity of L3T4-positive, I-A-restricted T cell clones. J Immunol 136(1):348–56

Kharitonenkov A, Chen Z et al. (1997) A family of proteins that inhibit signaling through tyrosine kinase receptors. Nature 386(6621):181–6

Kiener PA, Lioubin MN et al. (1997) Co-ligation of the antigen and Fc receptors gives rise to the selective modulation of intracellular signaling in B cells. Regulation of the association of phosphatidylinositol 3-kinase and inositol 5'-phosphatase with the antigen receptor complex. J Biol Chem 272(6):3838–44

Kuroiwa A, Yamashita Y et al. (1998) Association of tyrosine phosphatases SHP-1 and SHP-2, inositol 5-phosphatase SHIP with gp49B1, and chromosomal assignment of the gene. Journal of Biological Chemistry 273(2):1070–4

Lanier L (1998) NK cell receptors. Annual Review Immunology 16:359–393

Latour S, Fridman WH et al. (1996) Identification, molecular cloning, biologic properties, and tissue distribution of a novel isoform of murine low-affinity IgG receptor homologous to human Fc gamma RIIB1. Journal of Immunology 157(1):189–97

Leibson PJ (1997) Signal transduction during NK cell activation: inside the mind of a killer. Immunity 6(6):655–61

Li T, Rawlings DJ et al. (1997) Constitutive membrane association potentiates activation of Bruton tyrosine kinase. Oncogene 15(12):1375–83

Li Z, Wahl MI et al. (1997) Phosphatidylinositol 3-kinase-gamma activates Bruton's tyrosine kinase in concert with Src family kinases. Proceedings of the National Academy of Sciences of the United States of America 94(25):13820–5

Marengere LE, Waterhouse P et al. (1996) Regulation of T cell receptor signaling by tyrosine phosphatase SYP association with CTLA-4 [published errata appear in Science 1996 Dec 6;274(5293)1597 and 1997 Apr 4;276(5309):21]. Science 272(5265):1170–3

Muta T, Kurosaki T et al. (1994) A 13-amino-acid motif in the cytoplasmic domain of FcγRIIB modulates B-cell receptor signaling. Nature 368:70–73

Nadler MJS, Chen B et al. (1997) Protein-tyrosine phosphatase SHP-1 is dispensable for Fc gamma RIIB-mediated inhibition of B cell antigen receptor activation. Journal of Biological Chemistry 272(32): 20038–43

Nakamura K, Cambier JC (1998) B cell antigen receptor (BCR)-mediated formation of a SHP-2-pp120 complex and its inhibition by Fc gamma receptor IIB1 BCR coligation. The Journal of Immunology 161

O'Rourke LM, Tooze R et al. (1998) CD19 as a membrane-anchored adaptor protein of B lymphocytes: costimulation of lipid and protein kinases by recruitment of Vav. Immunity 8:635–645

Ono M, Bolland S et al. (1996) Role of the inositol phosphatase SHIP in negative regulation of the immune system by the receptor Fc gamma RIIB. Nature 383(6597):263–6

Ono M, Okada H et al. (1997) Deletion of SHIP or SHP-1 reveals two distinct pathways for inhibitory signaling. Cell 90:293–301

Pesando JM, Bouchard LS et al. (1989) CD19 is functionally and physically associated with surface immunoglobulin. Journal of Experimental Medicine 170:2159–2163

Phillips NE, Parker DC (1983) Fc-dependent inhibition of mouse B cell activation by whole anti-μ antibodies. Journal of Immunology 130:602–606

Phillips NE, Parker DC (1983) Fc-dependent inhibition of mouse B cell activation by whole anti-μ antibodies. J Immunol 130(2):602–6

Phillips NE, Parker DC (1984) Cross-linking of B lymphocyte Fc gamma receptors and membrane immunoglobulin inhibits anti-immunoglobulin-induced blastogenesis. Journal of Immunology 132(2):627–32

Pleiman CM, Hertz WM et al. (1994) Activation of Phosphatidylinositol-3' Kinase by Src-Family Kinase SH3 Binding to the p85 Subunit. Science 263:1609–1612

Pradhan M, Coggeshall KM (1997) Activation-induced bidentate interaction of SHIP and Shc in B lymphocytes. Journal of Cellular Biochemistry 67(1):32–42

Rigley KP, Harnett MM et al. (1989) Analysis of signaling via surface immunoglobulin receptors on B cells from CBA/N mice. European Journal of Immunology 19(11):2081–6

Salim K, Bottomley MJ et al. (1996) Distinct specificity in the recognition of phosphoinositides by the pleckstrin homology domains of dynamin and Bruton's tyrosine kinase. EMBO Journal 15:6241–6250

Sarkar S, Schlottmann K et al. (1996) Negative signaling via FcγRIIB1 in B cells blocks phospholipase Cγ2 tyrosine phosphorylation but not syk or lyn activation. Journal of Biological Chemistry 271:20182–20186

Satterthwaite AB, Cheroutre H et al. (1997) Btk dosage determines sensitivity to B cell antigen receptor cross-linking. Proceedings of the National Academy of Sciences of the United States of America 94(24):13152–7

Scharenberg A, El-Hillal O et al. (1998) Phosphatidylinositol-3,4,5-trisphosphate (PtdIns-3,4,5-P3)/Tec kinase-dependent calcium signaling pathway: a target for SHIP-mediated inhibitory signals. EMBO Journal 17(7):1961–1972

Sidorenko S, Law C-L et al. (1995) Human spleen tyrosine kinase p72Syk associates with the Src-family kinase p53/p56Lyn and a 120-kDa phosphoprotein. Proceedings of the National Academy of Sciences, USA 92:359–363

Sillman AL, Monroe JG (1995) Association of p72syk with the src homology-2 (SH2) domains of PLC gamma 1 in B lymphocytes. Journal of Biological Chemistry 270(20):11806–11

Takai T, Ono M et al. (1996) Augmented humoral and anaphylactic responses in FcγRII-deficient mice. Nature 379:346–349

Takata M, Kurosaki T (1996) A role for Bruton's tyrosine kinase in B cell antigen receptor-mediated activation of phospholipase C-gamma 2. Journal of Experimental Medicine 184(1):31–40

Tedder TF, Zhou LJ et al. (1994) The CD19/CD21 signal transduction complex of B lymphocytes. [Review]. Immunology Today 15(9):437–442

Thompson CB, Allison JP (1997) The emerging role of CTLA-4 as an immune attenuator. Immunity 7(4):445–50

Tridandapani S, Chacko GW et al. (1997) Negative signaling in B cells causes reduced Ras activity by reducing Shc-Grb2 interactions. J Immunol 158(3):1125–32

Tridandapani S, Kelley T et al. (1997) Recruitment and phosphorylation of SH2-containing inositol phosphatase and Shc to the B-cell Fc gamma immunoreceptor tyrosine-based inhibition motif peptide motif. Mol Cell 17(8):4305–11

Tuveson DA, Carter RH et al. (1993) CD19 of B cells as a Surrogate Kinase Insert Region to Bind Phosphatidylinositol 3-kinase. Science 260:986–989

Uher F, Dickler HB (1986) Independent ligand occupancy and cross-linking of surface Ig and Fc gamma receptors downregulates B-lymphocyte function. Evaluation in various B-lymphocyte populations. Molecular Immunology 23(11):1177–81

van Noesel CJM, Lankester AC et al. (1993) The CR2/CD19 complex on human B cells contains the src-family kin Lyn. International Immunology 5:699–705

Vely F, Olivero S et al. (1997) Differential association of phosphatases with hematopoietic coreceptors bearing immunoreceptor tyrosine-based inhibition motifs. European Journal of Immunology 27(8):1994–2000

Vely F, Vivier E (1997) Conservation of structural features reveals the existence of a large family of inhibitory cell surface receptors and noninhibitory/activatory counterparts. Journal of Immunology 159(5):2075–7

Wahl MI, Fluckiger AC et al. (1997) Phosphorylation of two regulatory tyrosine residues in the activation of Bruton's tyrosine kinase via alternative receptors. Proceedings of the National Academy of Sciences of the United States of America 94(21):11526–33

Waldschmidt TJ, Vitetta ES (1985) The use of haptenated immunoglobulin molecules to induce tolerance in B cells from neonatal mice. Journal of Immunology 134(3):1436–41

Weng W-K, Jarvis L et al. (1994) Signaling through CD19 Activates Vav/Mitogen-activated Protein Kinase Pathway and Induces Formation of a CD19/Vav/Phosphatidylinositol 3-Kinase Complex in Human B cell Precursors. Journal of Biological Chemistry 269(51):32514–32521

Wilson HA, Greenblatt D et al. (1987) The B lymphocyte calcium response to anti-Ig is diminished by membrane immunoglobulin cross-linkage to the Fc gamma receptor. Journal of Immunology 138(6):1712–1718

Yamashita Y, Fukuta D et al. (1998) Genomic structures and chromosomal location of p91, a novel murine regulatory receptor family. Journal of Biochemistry 123(2):358–368

Yokoyama WM (1998) Natural killer cell receptors. Current Opinion in Immunology 10(3):298–305

Regulation of B Cell Antigen Receptor Signaling by the Lyn/CD22/SHP1 Pathway

R.J. Cornall[1], C.C. Goodnow[2], and J.G. Cyster[3]

1 Introduction

In a resting cell, the B cell antigen receptor (BCR) is in an equilibrium of phosphorylated and unphosphorylated states that is tuned by counteracting kinases and phosphatases. Contributing to this tuning are transmembrane molecules (e.g., CD19/CD21, CD22, FcγRIIB1) whose cytoplasmic domains act as necessary docks for the activated kinases and phosphatases. When antigens cluster and increase the local density of BCRs, the equilibrium state is lost, tyrosine phosphorylation of BCR-associated molecules CD79 α/β, Syk kinase, CD22, and CD19 is markedly increased and there is recruitment and activation of further signaling enzymes leading to multiple downstream changes in the cell. The magnitude and duration of the activating signal continues to be influenced, however, by the balance of kinase

[1] Nuffield Department of Medicine, Oxford University, John Radcliffe Hospital, Headington, Oxford OX3 9DU, UK
[2] Australian Cancer Research Foundation Genetics Laboratory, Medical Genome Centre, John Curtin School of Medical Research, The Australian National University, Mills Rd, PO Box 334, Canberra, ACT 2601, Australia
[3] Department of Microbiology and Immunology, University of California, San Francisco, CA 94143-0414, USA

and phosphatase activity and the availability of transmembrane docking molecules. In this review, recent experiments will be discussed that have shown how BCR-induced activation of the src family kinase Lyn initiates the constitutive and antigen-induced feedback inhibition of the BCR through the phosphorylation of CD22 and recruitment of the phosphatase SHP1 to the CD22/BCR complex.

2 The Functional Importance of Antigen Receptor Signaling

The quality, quantity and timing of BCR signaling determines the response and selection of B cells (GOODNOW et al. 1995). In immature B cells in the bone marrow, continuous signaling by the BCR in response to self-antigen triggers a spectrum of tolerance responses, depending on the self-antigen. Self-antigens that avidly crosslink BCRs cause an arrest in development followed by apoptosis or editing of the BCR genes, thus eliminating strongly self-reactive cells from the repertoire exported to the spleen (GOODNOW et al. 1995; HERTZ and NEMAZEE 1998). Repeated binding of lower valency self-antigen induces a less stringent process of functional tolerance called anergy, in which the B cells mature and leave the bone marrow but their migration is altered, their lifespan is shortened, and the cells are actively desensitized by feedback mechanisms that decrease cell surface expression of IgM receptors and diminish proximal BCR signaling (HEALY and GOODNOW 1998). It is possible that anergic B cells could be recruited into an immune response by foreign antigens having higher epitope valency than the anergizing self-antigen, as has been established in model systems (COOKE et al. 1994; BACHMANN and ZINKERNAGEL 1997), and that this is why the cells are not eliminated more vigorously. This is a functional expression of the need to tune the BCR signaling response in such a way that the ability to mount an effective immune response against pathogens is balanced against the risk of autoimmune disease. Self-reactive B cells that bind self-antigens at too low a concentration or avidity to trigger B cell deletion or anergy can persist in the spleen and lymph nodes in a naive and fully activatable state (GOODNOW et al. 1995). Presumably, many weakly self-reactive B cells fall into this category, in which the danger of autoimmunity is strongly outweighed by the potential to fight infections.

In mature B cells, acute BCR signaling to foreign antigens triggers activation and, in conjunction with signals from T cells or innate immune mediators, clonal expansion and differentiation into antibody-secreting plasma cells. At this stage, activation by antigens is enhanced by greater BCR cross-linking (DINTZIS et al. 1989) and is modulated by coreceptors such as CD21 (the C3d receptor) and FcγRIIB1 (the receptor for IgG immune complexes), that alter the quality and quantity of BCR signaling (reviewed by CYSTER and GOODNOW 1997; DAERON 1997; O'ROURKE et al. 1997).

Deciphering how the BCR is tuned so that an immature or mature B cell responds only to a certain amount of antigen receptor engagement is a difficult problem to study in the normal B cell repertoire due to the unknown specificity and antigen-exposure history of the majority of B cells. This problem is largely over-come when mice carrying immunoglobulin transgenes are used; and the introduc-tion of mutations in key signaling components onto the controlled genetic background of Ig transgenic mice has proved a fruitful way of analyzing the mechanism by which BCR signaling is tuned.

3 The Phosphatase SHP1 Establishes the Threshold for BCR Activation

The role of the cytosolic tyrosine phosphatase SHP1 as an inhibitor of the BCR was first identified in studies of SHP1-deficient Ig-transgenic B cells. The motheaten viable mutation (me^v), which encodes a deficiency in SHP1, was bred into mice expressing the well-characterized immunoglobulin transgene against hen egg lyso-zyme (Ig^{HEL}) (CYSTER and GOODNOW 1995). Homozygous SHP-deficient (me^v/me^v) Ig^{HEL} B cells had a higher basal signaling, which caused them to modulate IgM spontaneously and hypersecrete the anti-HEL IgM antibody, despite the absence of antigens. The me^v/me^v Ig^{HEL} B cells initiated greater and more rapid elevation of intracellular calcium in response to the hen egg lysozyme (HEL) antigen than wild-type Ig^{HEL} B cells and were eliminated rather than anergized by exposure to low valency soluble HEL (sHEL) expressed as a neo-self-antigen. Like anergic B cells, naive SHP1-deficient B cells with their spontaneously exaggerated signaling were excluded from B cell follicles by competition from wild-type naive B cells (SCHMIDT et al. 1998).

4 Lyn-deficient and SHP1-deficient B Cells Have a Similar Phenotype

The src family protein tyrosine kinase Lyn is expressed abundantly in B cells (BOLEN and BRUGGE 1997). A small amount of cellular Lyn is physically associated with the BCR and activated upon BCR stimulation (BURKHARDT et al. 1991; CAMPBELL and SEFTON 1992; YAMANASHI et al. 1991). Chicken B lymphoma cells lacking Lyn have delayed and diminished BCR-induced intracellular calcium release, indicating that Lyn has a role in this process (KUROSAKI et al. 1994; TAKATA et al. 1994). Mice homozygous for disruption of the Lyn locus (lyn−/−), however, display normal immature B cell development, but have decreased num-bers of mature peripheral B cells, greatly elevated serum IgM and IgA, and produce autoantibodies that cause autoimmune glomerulonephritis reminiscent of systemic lupus erythematosus (CHAN et al. 1997; HIBBS et al. 1995; NISHIZUMI et al. 1995).

Antibody responses to antigenic challenge are relatively normal in lyn−/− mice, and Lyn-deficient splenic B cells make exaggerated proliferative and ERK responses after BCR clustering and relatively normal but delayed patterns of anti-immunoglobulin-induced protein phosphorylation (CHAN et al. 1997; WANG et al. 1996).

Analysis of lyn−/− Ig^{HEL} B cells revealed a phenotype very similar to that of SHP1-deficient Ig^{HEL} B cells (CORNALL et al. 1998). In the absence of HEL autoantigen, lyn−/− Ig^{HEL} B cells spontaneously upregulated MHC class II expression, downregulated surface, IgM, and differentiated into 20 times more plasma cells that spontaneously secreted 20 times more anti-HEL IgM antibody into the serum. By contrast, in the continuous presence of circulating sHEL-antigen, plasma cell formation and anti-HEL antibody secretion were completely suppressed, regardless of the lyn genotype. Indeed, negative selection to self-antigen was exaggerated by Lyn deficiency since, as in SHP1-deficient cells, maturational arrest and deletion were induced by sHEL and mirrored the normal tolerance response to forms of the antigen that clustered BCRs more avidly, such as membrane HEL.

As in SHP1-deficient B cells, the antigen-induced calcium flux in lyn−/− B cells was exaggerated (CORNALL et al. 1998). However, unlike SHP1 deficient cells, lyn−/− cells did show a consistent delay in the initial phase of the response, implying that Lyn plays a role in the initiation of calcium mobilization as well as its inhibition.

The similarity between SHP1 and Lyn deficiency raised the interesting question of whether the two proteins could be operating in a single pathway. To address this possibility, we employed an approach that has been very useful for analyzing signaling networks in lower eukaryotes by looking for genetic interaction between heterozygous mutants of Lyn and SHP1 when the dosage of functional genes was halved. We observed that B cells in heterozygous mice with a single functional lyn or SHP1 allele (lyn+/− or me^v/+) exhibited a degree of spontaneous IgM downregulation that lay between that of wild-type and lyn−/− or me^v/me^v-cells. This implied that a partial reduction in the function of either protein was detected within the B cells and resulted in a compensatory adjustment of surface IgM density. By intercrossing heterozygotes for the me^v mutation and the lyn− allele, double heterozygous Ig^{HEL} mice with partial deficiency in both proteins were produced. Simultaneous heterozygosity at both lyn and me^v loci caused B cells to adjust their surface IgM density to a lower level than heterozygosity at either locus alone, indicating a genetic interaction (ANOVA, $p = 0.0002$) (CORNALL et al. 1998).

5 Regulation of BCR Signaling by Lyn and SHP1 Occurs Via CD22

It has now been shown that Lyn is required for the phosphorylation of CD22 and the recruitment of SHP1 to the BCR (CORNALL et al. 1998; NISHIZUMI et al. 1998; SMITH et al. 1998). This observation extends earlier demonstrations that SHP1 is recruited to phosphorylated tyrosine-based inhibition motifs (ITIMs) on CD22 after BCR clustering (CAMPBELL and KLINMAN 1995; DOODY et al. 1995; LAW et al. 1996) and that CD22 and Lyn can be coimmunoprecipitated (TUSCANO et al. 1996). Like lyn−/− and SHP1-deficient B cells, CD22-deficient (cd22−/−) B cells show spontaneous downregulation of IgM and exaggerated calcium responses to BCR clustering (NITSCHKE et al. 1997; O'KEEFE et al. 1996; OTIPOBY et al. 1996; SATO et al. 1996). Significantly, Lyn determines not only the antigen-induced phosphorylation of CD22, but also the basal level of CD22 phosphorylation and SHP1 recruitment in naive B cells (CORNALL et al. 1998). Together with evidence that Lyn can associate with the resting BCR (PLEIMAN et al. 1993), this provides a basis for constitutive regulation of the BCR, since 0.2–2% of the surface IgM receptor on resting B cells is associated with CD22 (LEPRINCE et al. 1993; PEAKER and NEUBERGER 1993).

Lyn kinase is not essential for basal or antigen-induced phosphorylation of CD79α/β or Syk, although it appears to assist in the initial antigen-induced signaling because phosphorylation of CD79 α/β and Syk is diminished (CHAN et al. 1997; CORNALL et al. 1998) and the elevation of calcium occurs more slowly in HEL-stimulated lyn−/− Ig^{HEL} cells (CORNALL et al. 1998). It is likely that the other src kinases expressed in primary B cells, such as Fyn and Blk, compensate for Lyn in promoting these signaling steps because HEL-induced calcium, ERK signaling, and negative selection responses are diminished in Ig^{HEL} B cells lacking the src family activator CD45 (CYSTER et al. 1996). In CD45-deficient B cells, decreased activity of all the src kinase members would be expected to preclude compensation amongst them. The conclusion that other src kinases compensate for Lyn in promoting certain BCR signaling pathways is consistent with analyses of lyn−/− DT40 chicken B lymphoma cells. Lyn is the sole src kinase expressed in DT40 cells, and BCR signaling is delayed and diminished when these cells lack a functional lyn gene, much as in CD45−/− B cells (KUROSAKI et al. 1994; TAKATA et al. 1994). Transfection of lyn−/− DT40 cells with Lck or Fyn kinase restores BCR signaling and results in a delayed but exaggerated calcium response comparable to that of lyn−/− primary B cells.

To further examine the functional significance of the biochemical pathway involving Lyn, CD22, and SHP1 in the regulation of the BCR, we asked whether partial deficiency in CD22 could exaggerate the effects of partial Lyn and SHP1 deficiencies. Mice heterozygous for a targeted deletion of CD22 (cd22+/−) were mated with me^v/+ and lyn+/− Ig^{HEL} transgenic mice to generate sibships with fully wild-type or heterozygous mutant alleles at each of the three loci. This work showed that all three known elements of the regulatory pathway are limiting, since the presence of a single functional copy of any gene caused compensatory adjustment of surface IgM receptor density, in the order me^v > lyn > CD22 (CORNALL et al. 1998). Combinations of these partial deficiencies had cumulative effects on IgM receptor density and threshold effects on other responses to BCR signaling such as MHC class II upregulation, spontaneous antibody hypersecretion, and negative selection. The sensitivity to gene dosage and genetic interaction among the members of the pathway provide an example of polygenic control of both continuous (qualitative) and threshold (quantitative) traits.

The key features of the Lyn/SHP1/CD22 pathway are summarized in Fig. 1.

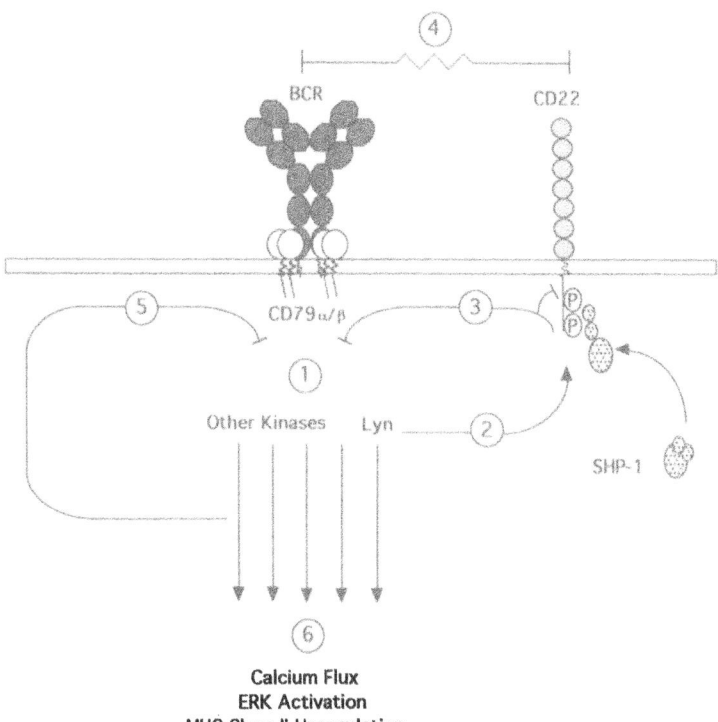

Fig. 1. Model illustrating the roles of Lyn, SHP-1, and CD22 in regulating BCR signaling and selection. *1*, Multiple tyrosine kinases promote phosphorylation of the CD79 α/β (Ig α/β) subunits of the BCR and signaling to second messengers such as calcium and ERK. This occurs at a low level in the absence of antigen and at a higher level when antigen-induced clustering increases the local concentration of BCRs, or when opposing tyrosine phosphatases are inhibited with pervanadate. *2*, Lyn kinase is essential for basal and antigen-induced phosphorylation of CD22 on ITIM tyrosine residues. *3*, SHP-1 binds via its SH2 domains to the phosphorylated ITIMs on CD22 and thus is recruited and activated near the BCR, where it inhibits signaling by dephosphorylating as yet undefined substrates, which may include CD22. *4*, Interaction between CD22 and external ligands may adjust the proximity of CD22 to the BCR and modify the extent of SHP-1-mediated regulation. *5*, IgM modulation in response to increased signaling reduces receptor density and provides a separate feedback inhibition of BCR signaling. *6*, Signaling to calcium, ERK, and other second messengers is quantitatively and qualitatively altered by different combinations of lyn, CD22, and SHP-1 alleles, resulting in B cell hyperactivity

6 The Implications of a Pathway Comprising Several Limiting Components

Three implications arise from the demonstration that each component in the BCR regulatory pathway is limiting. Firstly, it is clear that the genetic enhancer/suppressor approaches used to understand signaling networks in Drosophila or *C. elegans* (KARIM et al. 1996; SOMMER and STERNBERG 1996; VERHEYEN et al.

1996; WASSARMAN et al. 1995) should also be possible in mice, and that existing assays, such as IgM modulation, can be used to further extend the analysis of BCR signaling. Secondly, the cumulative effects of multiple weak alleles in the manner described for the Lyn/SHP1/CD22 pathway characterize genetic predisposition to many complex diseases such as autoimmune diabetes and systemic lupus erythematosus (VYSE and TODD 1996). In these diseases, multiple loci contribute to the relative risk of disease, and locus heterogeneity exists such that different combination of alleles can have equivalent effects on the risk of disease. Elements that tune the immune response and are limiting at the level of single alleles are highly likely to be subject to natural selection and become the site of inherited autoimmune susceptibility genes. Thirdly, the physiological significance of dosage sensitivity at each step in this signaling pathway may be that processes activating, inhibiting, or sequestering these molecules can tightly regulate the supply of each protein.

Cross-linking CD22 independently of the BCR complex can enhance signaling from the receptor (PEZZUTTO et al. 1987; PEZZUTTO et al. 1988; TUSCANO et al. 1996), and it has been suggested that this may be due to the sequestration of SHP1. One function of the ectodomain of CD22 could be to regulate the extent to which the cytoplasmic tail is associated with the BCR (reviewed, CYSTER and GOODNOW 1997). CD22 contains a sialic acid α2-6 Gal binding lectin on its two amino-terminal domains and is theoretically capable of binding to a range of serum factors and immune receptors, as well as to sialyted molecules within the lymphoid microenvironment. In similar fashion, sequestration of Lyn and SHP1 may also modify the regulation of the BCR. It has recently been proposed that sequestration of a limiting supply of Lyn in mast cells might be at the basis of FcϵRI receptor antagonism (TORIGOE et al. 1998). Similar effects could occur in B cells where Lyn associates with several receptors in addition to the BCR and CD22, including CD19, CD20, and CD40.

7 Multiple Inhibitory Pathways May Exist in B Cells

The longstanding hypothesis that CD22 has positive regulatory functions which extend beyond the sequestration of SHP1 is supported by observation of the coprecipitation of Syk, Phosphatidylinositol-3-kinase (PI-3-kinase), and phospholipase Cγ(PLCγ) with CD22 (LAW et al. 1996; TUSCANO et al. 1996), and the presence of ITAMs as well as ITIMs on the coreceptor. Positive effects of CD22 might explain why lyn$-/-$ and me^v/me^v B cells have a more severe phenotype than the CD22$-/-$ B cells, which are not eliminated by low valency self-antigen (RJC, unpublished data). Alternatively, Lyn and SHP1 may have more marked phenotypes because they are involved in other regulatory pathways operating parallel to CD22. For example, Lyn phosphorylates the Sos inhibitor Cbl in DT40 cells (TE-ZUKA et al. 1996) and is also required to phosphorylate FcγRIIB1 in negative regulation of B cells during an immune response (NISHIZUMI et al. 1998); and SHP1 associates with Vav, Sos and Grb2 (KON-KOZLOWSKI et al. 1996).

Several new Ig-superfamily receptors have recently been described that bind SHP1 and inhibit the function of immune receptors (BORGES et al. 1997; WAGTMANN et al. 1997; reviewed CAMBIER 1997; YOKOYAMA 1988). Of these receptors, the murine paired Ig-like receptor-B (PIR-B) is expressed by B lymphocytes and in cross-linking experiments was able to inhibit the antigen-induced calcium response (BLERY et al. 1998; MAEDA et al. 1998). Like CD22, PIR-B is a member of the killer inhibitory receptor (KIR) family encoded on a single region of proximal mouse chromosome 7 (ALLEY et al. 1998). Human ILT2/LIR-1, the human homologue of PIR-B, binds MHC class I molecules (COLONNA et al. 1997; COSMAN et al. 1997). Similarly, the ligand of murine PIB-B may be a class I molecule, but this has yet to be confirmed.

The B cell molecule CD72, which carries a C-type lectin domain similar to that of NK receptors CD94/NKG2 and Ly-49, was also recently found to act as a target for Lyn in vitro and to recruit SHP1 to its ITIM domain after BCR engagement (ADACHI et al. 1998). This result is reminiscent of the findings with CD22 because previous experiments had shown that treatment with anti-CD72 augmented BCR signaling. So it is tempting to speculate that CD72 might tune the BCR to its environment in a manner similar to CD22. The only ligand so far described for CD72 is another SHP1 docking protein, CD5. The significance of a CD5-CD72 interaction is controversial, but since CD5 is present on all T cells, it is possible that CD72-mediated regulation of BCR signaling could be influenced by proximity to T cells. In vivo, BCR engagement causes B cells to move from follicles to T cell-rich areas of lymphoid tissues where extensive engagement of CD72 could occur with consequent adjustments in the strength of BCR signaling. Other ligands within the lymphoid microenvironment could have similar effects.

CD5 is also expressed on the B-1 subset of B cells, where it also can act as a negative regulator, presumably through recruitment of SHP1. Studies in CD5-deficient mice have established that CD5 negatively regulates the antigen-induced calcium response, nuclear translocation of NF-kB, and proliferative response of B1-B cells (BIKAH et al. 1996). SHP1 deficiency on the Ig^{HEL} background results in increased numbers of B-1 B cells (CYSTER and GOODNOW 1995), but this is not mirrored in lyn−/− Ig^{HEL} cells (CORNALL et al. 1998) and suggests that any SHP1-mediated inhibition of B-1 cell development does not require Lyn. Whether the negative regulatory function of CD5 in B1-B cells requires Lyn has not been determined.

8 What Are the Targets of the CD22- SHP1 Complex?

The downstream substrates of dephosphorylation by CD22-SHP1 are unknown. KIR engagement results in dephosphorylation of a subset of proteins involved in the early stages of signaling, including ZAP-70, PLCγ, and its adapter pp36 (BINSTADT et al. 1996); and similar proteins might be targets in B cells. The absence of detectably enhanced phosphorylation of CD79 α/β and Syk in unstimulated lyn−/−lyn Ig^{HEL} B cells suggests that these are not the targets (CORNALL et al. 1998), although any effects may well be outside the sensitivity of the assays we have performed so far.

Some indication of the possible targets of CD22-SHP1 is given by an analysis of the CD22 versus CD19 counter-regulation of BCR-induced MAP kinase activity (TOOZE et al. 1997). Coligation of CD22 alone enhances activation of ERK2, JNK, and p38, presumably by sequestering SHP1. In contrast, ligation of CD22 to the BCR inhibits activation of the same kinases induced by stimulation of both BCR and CD19. CD19 can have a constitutive as well as antigen-induced effect on BCR signaling, because overexpression of CD19 from a transgene augments BCR signaling and induces a B cell phenotype similar to that of SHP1 deficiency (INAOKI et al. 1997). These observations suggest that constitutive counter-regulation of CD19 and CD22 may be a physiological phenomenon in unstimulated B cells, as well as during coligation of the BCR and CD19-CD21 through C3d decorated antigen. The targets of CD19 activation may also be the targets of CD22 inhibition. Elegant studies of the relationships between CD19, Vav, PI-3-kinase, Btk, FcγRIIB1, and the phosphatase SHIP provide a theoretical basis for further research in this area (ONO et al. 1997; O'ROURKE et al. 1998; BOLLARD et al. 1998; and reviews by SCHARENBERG and KINET 1998; CANTRELL 1998). Identifying the targets of CD22-SHP1, perhaps by using substrate-trapping mutants of SHP1 (TONKS and NEEL 1996), remains an important area for future investigation.

9 Concluding Remarks

It is hardly surprising that the flexibility required for BCR signaling is achieved by a complex network of biochemical pathways, and that within this network the distinction between activation and inhibition at times seems somewhat arbitrary. Nevertheless, the notion that a balance of ITAM-bound kinases and ITIM-bound phosphatases establishes the threshold for BCR signaling has proved an extremely useful paradigm. We and others have demonstrated in B cells that these two events may be linked by the activity of the src-kinase Lyn through the phosphorylation of CD22 and recruitment of SHP1. These studies have given some insight into the genetic basis of autoimmune susceptibility and the nature and implications of rate limiting steps. For the immediate future, they have demonstrated that complex pathways may be reduced for functional and genetic analysis by models in which immune repertoire, and genetic and environmental effects are carefully controlled. The rapid pace of this integrated field gives hope that we may soon be in a position to develop new therapeutic reagents targeted at BCR modulation.

Acknowledgment. R.J.C. is a Wellcome Trust Clinician Scientist.

References

Adachi T, Flaswinkel H, Yakura H, Reth M, Tsubata T (1998) The B cell surface protein CD72 recruits the tyrosine phosphatase SHP-1 upon tyrosine phosphorylation. J Immunol 160:4662–4665

Alley TL, Cooper MD, Chen M, Kubagawa (1998) Genomic structure of PIR-B, the inhibitory member of the paired immunoglobulin-like receptor genes in mice. Tissue Antigens 51:224–231

Bachmann MF, Zinkernagel RM (1997) Neutralizing antiviral B cell responses. Annu Rev Immunol 15:235–270

Bikah G, Carey J, Ciallella JR, Tarakhovsky A, Bondad S (1996) CD5-mediated negative regulation of antigen receptor-induced growth signals in B-1 B cells. Science 274:1906–1909

Binstadt BA, Brumbaugh KM, Dick CJ, Scharenberg AM, Williams BL, Colonna M, Lanier LL, Kinet JP, Abraham RT, Leibson PJ (1996) Sequential involvement of Lck and SHP-1 with MHC-recognizing receptors on NK cells inhibits FcR-initiated tyrosine kinase activation. Immunity 5:629–638

Blery M, Kubagawa H, Chen C-C, Vely F, Cooper MD, Vivier E (1998) The paired Ig-like receptor PIR-B is an inhibitory receptor that recruits the protein-tyrosine phosphatase SHP-1. Proc Natl Acad Sci USA 95:2446–2451

Bolen JB, Brugge JS (1997) Leukocyte protein tyrosine kinases: potential targets for drug discovery. Annu Rev Immunol 15:371–404

Bollard S, Pearse RN, Kurosaki T, Ravetch JV (1998) SHIP modulates immune receptor responses by regulating membrane association of Btk. Immunity 8:509–516

Borges L, Hsu M-L, Fanger N, Kubin M, Cosman D (1997) A family of human lymphoid and myeloid receptors, some of which bind to MHC class I molecules. J Immunol 159:5192–5196

Burkhardt AL, Brunswick M, Bolen JB, Mond JJ (1991) Anti-immunoglobulin stimulation of B lymphocytes activates src-related protein-tyrosine kinases. Proc Natl Acad Sci USA 88:7410–7414

Cambier JC (1997) Inhibitory receptors abound? Proc Natl Acad Sci USA 94:5993–5995

Campbell M-A, Sefton BM (1992) Association between B-lymphocyte membrane immunoglobulin and multiple members of the src family of protein tyrosine kinases. Mol Cell Biol 12:2315–2321

Campbell MA, Klinman NR (1995) Phosphotyrosine-dependent association between CD22 and the protein tyrosine phosphatase 1C. Eur J Immunol 25:1573–1579

Cantrell D (1998) Lymphocyte signaling: a coordinating role for Vav? Curr Biol 8:535–538

Chan VWF, Meng F, Soriano P, DeFranco AL, Lowell CA (1997) Characterization of the B lymphocyte populations in Lyn-deficient mice and the role of Lyn in signal initiation and downregulation. Immunity 7:69–81

Colonna M, Navarro F, Bellon T, Llano M, Garcia P, Samaridis J, Angman L, Cella M, Lopez-Botet M (1997) A common inhibitory receptor for major histocompatibility complex class I molecules on human lymphoid and myelomonocytic cells. J Exp Med 186:1809–1818

Cooke MP, Heath AW, Shokat KM, Zeng Y, Finkelman FD, Linsley PS, Howard M, Goodnow CC (1994) Immunoglobulin signal transduction guides the specificity of B cell/T cell interactions and is blocked in tolerant self-reactive B cells. J Exp Med 179:425–438

Cornall RJ, Cyster JG, Hibbs ML, Dunn AR, Otipoby KL, Clark EA, Goodnow CC (1998). Polygenic autoimmune traits: Lyn kinase, CD22, and SHP-1 phosphatase are limiting elements of a biochemical pathway that regulates BCR signaling and selection. Immunity 8:497–508

Cosman DN, Fanger N, Borges L, Kubin M, Chin W, Peterson L, HSU M-L (1997) A novel immunoglobulin superfamily receptor for cellular and viral MHC class I molecules. Immunity 7:273–282

Cyster JG, Goodnow CC (1995) Protein tyrosine phosphatase 1C negatively regulates antigen receptor signaling in B lymphocytes and determines thresholds for negative selection. Immunity 2:13–24

Cyster JG, Goodnow CC (1997) Tuning antigen receptor signaling by CD22: integrating cues from antigens and the microenvironment. Immunity 6:509–517

Cyster JG, Healy JI, Kishihara K, Mak TW, Thomas ML, Goodnow CC (1996) CD45 sets thresholds for negative and positive selection of B lymphocytes. Nature 381:325–8

Daeron M (1997) Fc receptor biology. Annu Rev Immunol 15:203–234

Dintzis RZ, Okajima M, Middleton MH, Greene G, Dintzis HM (1989) The immunogenicity of soluble haptenated polymers is determined by molecular mass and hapten valence. J Immunol 143:1239–1244

Doody GM, Justement LB, Delibrias CC, Matthews RJ, Lin J, Thomas ML, Fearon DT (1995) A role in B cell activation for CD22 and the protein tyrosine phosphatase SHP. Science 269:242–244

Goodnow C, Cyster J, Hartley S, Bell S, Cooke M, Healy J, Akkaraju S, Rathmell J, Pogue S, Shokat K (1995) Self-tolerance Checkpoints in B Lymphocyte Development. Adv Immunol 59:279–368

Healy JI, Goodnow CC (1998) Positive versus negative signaling by lymphocyte antigen receptors. Annu Rev Immunol 16:645–670

Hertz M, Nemazee D (1998) Receptor editing and commitment in B lymphocytes. Curr Opin Immunol 10:208–213

Hibbs ML, Tarlinton DM, Armes J, Grail D, Hodgson G, Maglitto R, Stacker SA, Dunn AR (1995) Multiple defects in the immune system of lyn-deficient mice, culminating in autoimmune disease. Cell 83:301–311

Inaoki M, Sato S, Weintraub BC, Goodnow CC, Tedder TF (1997) CD19-regulated signaling thresholds control peripheral tolerance and autoantibody production in B lymphocytes. J Exp Med 186: 1923–1931

Karim FD, Chang HC, Therrien M, Wassarman DA, Laverty T, Rubin GM (1996) A screen for genes that function downstream of Ras1 during drosophila eye development. Genetics 143:315–329

Kon-Kozlowski M, Pani G, Pawson T, Siminovich KA (1996) The tyrosine phosphatase PTP1C associates with Vav, Grb2, and mSos1 in hematopoietic cells. J Biol Chem 271:3856–3862

Kurosaki T, Takata M, Yamanashi Y, Inazu T, Taniguchi T, Yamamoto T, Yamamura H (1994) Syk activation by the src-family tyrosine kinase in B cell receptor signaling. J Exp Med 179:1725–1729

Law C-L, Sidorenko SP, Chandran KA, Zhao Z, Shen S-H, Fischer EH, Clark EA (1996) CD22 associates with protein tyrosine phosphatase 1C, Syk, and phospholipase C-g1 upon B cell activation. J Exp Med 183:547–560

Leprince C, Draves KE, Geahlen RL, Ledbetter JA, Clark EA (1993) CD22 associates with the human surface IgM-B-cell antigen receptor complex. Proc Natl Acad Sci USA 90:3236–3240

Maeda A, Kurosaki M, Ono M, Takai T, Kurosaki T (1998) Requirement of SH2-containing protein tyrosine phosphatases SHP-1 and SHP-2 for paired immunoglobulin-like receptor B (PIR-B)-mediated inhibitory signal. J Exp Med 187:1355–1360

Nishizumi H, Horikawa K, Mlinaric-Rascan I, Yamamoto T (1998) A double-edged kinase Lyn: a positive and negative regulator for antigen receptor-mediated signals. J Exp Med 187:1343–1348

Nishizumi H, Taniuchi I, Yamanashi Y, Kitamura D, Ilic D, Mori S, Watanabe T, Yamamoto T (1995) Impaired proliferation of peripheral B cells and indication of autoimmune disease in lyn-deficient mice. Immunity 3:549–560

Nitschke L, Carsetti R, Ocker B, Kohler G, Lamers, MC (1997) CD22 is a negative regulator of B cell receptor signaling. Curr Biol 7:133–143

O'Keefe TL, Williams GT, Davies SL, Neuberger MS (1996) Hyperresponsive B cells in CD22-deficient mice. Science 274:798–801

O'Rourke L, Tooze R, Fearon DT (1997) Co-receptors of B lymphocytes. Curr Opin Immunol 9:324–329

O'Rourke LM, Tooze R, Turner M, Sandoval DM, Carter RH, Tybulewicz VLJ, Fearon DT (1998) CD19 as a membrane-anchored adaptor protein of B lymphocytes: costimulation of lipid and protein kinases by recruitment of Vav. Immunity 8:635–645

Ono M, Okada H, Bolland S, Yanagi S, Kurosaki T, Ravetch JV (1997) Deletion of SHIP or SHP-1 reveals two distinct pathways for inhibitory signaling. Cell 90:293–301

Otipoby KL, Andersson KB, Draves KE, Klaus SJ, Garr AG, Kerner J D, Perlmutter RM, Law C-L, Clark EA (1996) CD22 regulates thymus-independent responses and the lifespan of B cells. Nature 384:634–637

Peaker CJ, Neuberger MS (1993) Association of CD22 with the B cell antigen receptor. Eur J Immunol 23:1358–1363

Pezzutto A, Dorken B, Moldenhauer G, Clark EA (1987) Amplification of human B cell activation by a monoclonal antibody to the B cell-specific antigen CD22, Bp 130/140. J Immunol 138: 98–103

Pezzutto A, Rabinovitch PS, Dorken B, Moldenhauer G, Clark EA (1988) Role of the CD22 human B cell antigen in B cell triggering by anti-immunoglobulin. J Immunol 140:1791–1795

Pleiman CM, Clark MR, Timson Gauen LK, Winitz S, Coggeshall KM, Johnson GL, Shaw AS, Cambier JC (1993) Mapping of sites on the src family protein tyrosine kinases p55^{blk}, p59^{fyn}, and p56^{lyn} which interact with the effector molecules phospholipase C-γ2, microtubule-associated protein kinase, GTPase-activating protein, and phosphatidylinositol 3-kinase. Mol Cell Biol 13:5877–5887

Sato S, Miller AS, Inaoki M, Bock CB, Jansen PJ, Tang ML, Tedder TF (1996) CD22 is both a positive and a negative regulator of B lymphocyte antigen receptor signal transduction: altered signaling in CD22-deficient mice. Immunity 5:551–562

Scharenberg AM, Kinet JP (1998) PtdIns-3,4,5-P3: a regulatory nexus between tyrosine kinases and sustained calcium signals. Cell 94:5–8

68   R.J. Cornall et al.: Regulation of B Cell Antigen Receptor Signaling

Schmidt KN, Hsu CW, Griffin CT, Goodnow CC, Cyster JG (1998) Spontaneous follicular exclusion of SHP1-deficient B cells is conditional on the presence of competitor wild-type B cells. J Exp Med 187: 929–937

Smith KGC, Tarlinton DM, Doody GM, Hibbs ML, Fearon DT (1998) Inhibition of the B cell by CD22: a requirement for Lyn. J Exp Med 187:807–811

Sommer RJ, Sternberg PW (1996) Apoptosis and change of competence limit the size of the vulva equivalence group in Pristionchus pacificus: a genetic analysis. Curr Biol 6:52–59

Takata M, Sabe H, Hata A, Inazu T, Homma Y, Nukada T, Yamamura H, Kurosaki T (1994) Tyrosine kinases Lyn and Syk regulate the B cell receptor-coupled Ca^{2+} mobilization through distinct pathways. EMBO J 13:1341–1349

Tezuka T, Umemori H, Fusaki N, Yagi T, Takata M, Kurosaki T, Yamamoto T (1996) Physical and functional association of the cbl protooncogene with a src-family protein tyrosine kinase, $p53/56^{lyn}$, in B cell antigen receptor signaling. J Exp Med 183:675–680

Tonks NK, Neel BG (1996) From form to function: signaling by protein tyrosine phosphatases. Cell 87:365–368

Tooze RM, Doody GM, Fearon DT (1997) Conterregulation by the coreceptors CD19 and CD22 of MAP kinase activation by membrane immunoglobulin. Immunity 7:59–67

Torigoe C, Inman JK, Metzger H (1998) An unusual mechanism for ligand antagonism. Science 281: 568–572

Tuscano J, Engel P, Tedder TF, Agarwal A, Kehrl JH (1996) Involvement of p72syk kinase, p53/56lyn kinase and phosphatidyl inositol-3 kinase in signal transduction via the human B lymphocyte antigen CD22. Eur J Immunol 26:1246–1252

Tuscano J, Engel P, Tedder TF, Kehrl JH (1996). Engagement of the adhesion receptor CD22 triggers a potent stimulatory signal for B cells and blocking CD22/CD22L interactions impairs T-cell proliferation. Blood 87:4723–4730

Verheyen EM, Purcell KJ, Fortinin ME, Artavanis-Tsakonas S (1996) Analysis of dominant enhancers and suppressors of activated Notch in Drosophila. Genetics 144:1127–1141

Vyse TJ, Todd JA (1996) Genetic analysis of autoimmune disease. Cell 85:311–318

Wagtmann N, Rojo S, Eichler E, Mohrenweiser H, Long EO (1997) A new gene complex encoding the killer cell inhibitory receptors and related monocyte/macrophage receptors. Curr Biol 7:615–618

Wang J, Koizumi T, Watanabe T (1996) Altered antigen receptor signaling and impaired Fas-mediated apoptosis of B cells in Lyn-deficient mice. J Exp Med 184:831–838

Wassarman DA, Therrien M, Rubin GM (1995) The Ras signaling pathway in Drosophila. Curr Opin Genet Dev 5:44–50

Yamanashi Y, Kakiuchi T, Mizuguchi J, Yamamoto T, Toyoshima K (1991) Association of B cell antigen receptor with protein tyrosine kinase Lyn. Science 251:192–194

Yokoyama WM (1988) What goes up must come down: the emerging spectrum of inhibitory receptors. J Exp Med 186:1803–1808

HLA-Specific and Non-HLA-Specific Human NK Receptors

A. Moretta[1], C. Bottino[2], R. Millo[1], and R. Biassoni[3]

1 Introduction

Unlike cytolytic T cells, which recognize and kill target cells expressing MHC molecules, NK lymphocytes can efficiently lyse target cells that lack the expression of one or more MHC class I molecules. Presumably, this allows the immune system to detect and kill certain types of tumor or virus-infected cells trying to evade T cell recognition by downregulating one or more class I molecules (GARRIDO et al. 1995; IKEDA et al. 1997). Based on the above observation, the cytolytic activity mediated by NK cells was originally viewed as non-MHC-restricted (KIESSLING et al. 1975; HERBERMAN et al. 1975; TRINCHIERI 1989). Although it is likely that triggering receptor(s) involved in the induction of NK-cell-mediated cytotoxicity are indeed represented by structures that recognize non-MHC ligands, the importance of MHC molecules in controlling NK functions is now well-established. Thus, it was shown that NK cells have the ability to kill "normal nonself" cells while sparing "normal self" cells expressing normal levels of MHC class I molecules and at least some allogeneic target cells expressing MHC class I molecules that are identical or related to the self ones (KARRE 1992; MORETTA et al. 1994; MORETTA 1992). In humans, this is possible thanks to the expression, on NK cells, of a series of

[1] Dipartimento di Medicina Sperimentale, University of Genoa, Italy
[2] Istituto Scientifico Tumori, Centro di Biotecnologie Avanzate, Genoa, Italy
[3] Istituto Scientifico Tumori, Centro di Biotecnologie Avanzate, Genoa, Italy

receptors that recognize HLA class I molecules (MORETTA et al. 1996; VALIANTE et al. 1997). Some of these, by delivering inhibitory signals to the NK cells, avoid eliminating normal self cells. Other HLA class-I-specific receptors exist that induce NK cell triggering and might be involved in the elimination of allogeneic cells or autologous cells expressing "abnormal" HLA molecules. Moreover, since NK cells are also able to kill HLA-negative target cells, it is evident that activating receptors also exist that trigger NK cells upon recognition of non-HLA ligands. The complex and variable interaction of all the different types of NK receptors with their HLA or non-HLA ligands is responsible for the fine tuning and regulation of NK-cell-mediated cytotoxicity against normal vs transformed cells.

The present review will be focused on the NK receptors that are expressed on human NK cells.

2 Inhibitory NK Receptors Belonging to the Ig Superfamily

The first family of inhibitory NK receptors that has been identified was specific for HLA-C molecules (MORETTA et al. 1993). These receptors, termed p58 (58kDa; MORETTA et al. 1996), were expressed only by subsets of NK cells, thus suggesting that only a fraction of NK cells may recognize HLA-C. The p58 receptors could be divided into two different subgroups termed p58.1 and p58.2 on the basis of the reactivity with different monoclonal antibodies and the basis of their specificity for distinct groups of HLA-C alleles. The p58.1 receptor (now termed CD158a; MORETTA et al. 1997c) recognized by the EB6 mAb (MORETTA et al. 1990a) and the p58.2 receptor (CD158b) recognized by the GL183 mAb (MORETTA et al. 1990b) were found to be specific for a dimorphism in residues 77–80 in the pocket F of the α1 domain of HLA-C molecules. In particular, the HLA-C alleles recognized by the p58.1 receptor share the Asn 77 and Lys 80, whereas those recognized by the p58.2 receptors are characterized by Ser 77 and Asn 80 (COLONNA et al. 1993; BIASSONI et al. 1995; CICCONE et al. 1992a).

The interaction between p58 receptors and their HLA-C ligands leads to the inhibition of NK cell-mediated cytotoxicity that results in the protection of target cells expressing the correct HLA-C alleles (MORETTA et al. 1993). NK clones can coexpress both p58.1 and p58.2 receptors and, as a result, only target cells lacking the expression of HLA-C molecules are susceptible to lysis (VITALE et al. 1995). The specificity of the p58 receptors for HLA-C molecules was originally demonstrated by experiments in which the cytolytic activity of NK cell clones against HLA negative target cells transfected with HLA-C alleles could be restored in the presence of anti-p58 mAb (MORETTA et al. 1993). In these experimental conditions, the masking of the inhibitory receptor prevented the HLA-C/P58 receptor interaction and subsequent delivery of the inhibitory signals. The subsequent molecular cloning of the p58 molecules (WAGTMANN et al. 1995; COLONNA and SAMARIDIS 1995) indicated that these receptors are type I transmembrane proteins belonging to the

Ig superfamily, characterized by an extracellular region of 224 amino acid with two C2 type Ig-like domains, a transmembrane portion of 20 nonpolar amino acid, and a cytoplasmic region variable in length (76 or 84 amino acid) (Fig. 1). The ectodomains contain 2–4 N-linked glycosilation sites. Analysis by Southern blot indicated that the genes encoding for p58 receptors do not undergo somatic gene rearrangement and that the different p58 transcripts are encoded by a multigene family localized on human chromosome 19. The cytoplasmic tail of the p58 receptors is characterized by the presence of two V/IxYxxL/V sequences, termed immunoreceptor tyrosine-based inhibiting motif (ITIM), that are directly involved in negative signaling. Indeed, upon receptor cross-linking, these motifs are tyrosine phosphorylated by the associated p56^{lck} (BOTTINO et al. 1994; BINSTADT et al. 1996) and bind to the SH2-containing tyrosine phosphatases SHP-1 and SHP-2 (BURSHTYN et al. 1996; OLCESE et al. 1996; CAMPBELL et al. 1996). These phosphatases mediate dephosphorylation of adaptor/effector molecules that are part of the NK cell-activating signaling pathway (VALIANTE et al. 1996; WEBER et al. 1998). Similar results were reported with another member of the ITIM-bearing receptor family termed FcγRIIB; however, the p58 receptors, unlike FcγRIIB, do not bind to SHIP (polyphosphate inositol 5-phosphatase; VELY et al. 1997). These data indicate that the differential recruitment of phosphatases by distinct ITIM-bearing receptors may reflect diverse strategies of inhibition necessary to suppress different ITAM-bearing triggering molecules.

Although only a fraction of the total NK cell pool express one or another member of the p58 family of receptors, the p58-negative subset also recognize HLA class I molecules (CICCONE et al. 1992b). This was demonstrated in a series of

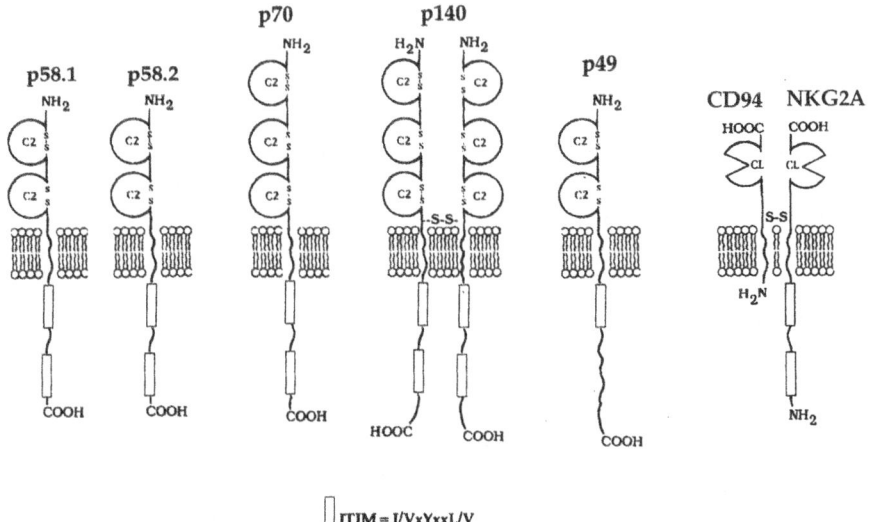

Fig. 1. HLA-specific inhibitory NK receptors

experiments in which virtually all p58-negative NK cell clones were inhibited by the expression of HLA-class I molecules on target cells and their cytolytic activity could be reversed by masking the protective HLA alleles (CICCONE et al. 1994). Thus, p58-negative NK clones were used for mice immunization to obtain new mAbs specific for other, still unknown, receptors for HLA class I molecules. The receptors for HLA-B and for HLA-A alleles were identified. In particular, two mAbs termed DX9 (LITWIN et al. 1994; GUMPERZ et al. 1995) and Z27 (VITALE et al. 1996) could be isolated that recognized a 70-kDa (p70) molecule expressed by human NK clones specific for HLA-B alleles displaying Bw4 supertypic specificity. These mAbs could reverse the cytolytic activity of the NK clones expressing the p70 receptor but not that of p58+ clones specific for HLA-C. The genes encoding the p70 NK receptor are also located on chromosome 19 (COLONNA and SAMARIDIS 1995; D'ANDREA et al. 1995; CANTONI et al. 1996) and encode type I transmembrane proteins belonging to the Ig superfamily and display a high degree of sequence homology with the p58 receptors. The most remarkable difference is that the p70 receptors contain three Ig-like extracellular domains instead of two. The (84 amino acid) cytoplasmic region of the p70 receptor contains two ITIMs involved in the suppression of NK-mediated cytotoxicity (FRY et al. 1996; Fig. 1).

Analysis of the role of HLA-bound peptides in recognizing HLA-B by p70+ NK clones originally indicated that only some of the peptides bound to the HLA-B*2705 molecule are able to have a protective effect (MALNATI et al. 1995). Moreover, the analysis of a panel of peptides corresponding to known endogenous ligands for HLA-B*2705 indicated that peptides unrelated by length and sequence (except for the anchor residue at position 2) were efficiently recognized by the p70 receptor. Importantly, substitution experiments demonstrated that side chains at positions 7 and 8 were crucial for recognition (PERUZZI et al. 1996). The side chain at position 8 of a nine-amino acid peptide bound to B*2705 was demonstrated to contact residues 76, 77, and 80 of the HLA class I heavy chain (MADDEN et al. 1992). These residues (see above) are part of the region that controls NK recognition both by p70+ and p58+ NK clones.

Peptide selectivity also exists in the recognition of HLA-Cw4 by p58.1 receptors, and positions 7 and 8 are crucial for recognition by these receptors (RAJAGOPALAN and LONG 1997). Additional studies indicated that the p58.2/Cw3 interaction also displays some peptide selectivity (ZAPPACOSTA et al. 1997). However, so far, the physiological relevance of this peptide selectivity is not well understood. Indeed, it is difficult to figure out a physiological situation causing a drastic change in the pool of peptides able to bind to a given HLA-class I allele such that target cells would loose their resistance to NK cells bearing HLA class-I-specific receptors.

Later on, mice immunization with p58/p70 negative clones characterized by the ability to recognize some HLA-A alleles (including A3 and A11) led to the isolation of mAbs, termed Q66 and Q241, specific for an HLA-A receptor. This receptor displays a molecular mass of 140kDa, and its masking by specific mAbs was able to reconstitute NK-mediated lysis of target cell transfectants expressing one or another of the protective HLA-A alleles (PENDE et al. 1996). By the use of a different experimental approach, another laboratory obtained a mAb, termed 5.133, that recognized both the HLA-A receptor recognized by Q66 and Q241 mAbs and the p70 receptor for HLA-Bw4 alleles (DÖHRING et al. 1996). Molecular cloning revealed that the p140 genes are located on human chromosome 19 and encode a protein belonging to the Ig superfamily displaying a three Ig-like domain structure that shares a high degree of sequence homology with p58 and p70 receptors. Comparative analysis indicated that p140 differ from the p70 receptor by approximately 50 amino acids in the extracellular portion. Thus, two Cys at position 302 and 336 proximal to the transmembrane portion are present in the p140 but not in the p70 receptors (PENDE et al. 1996). At least one of these Cys may be involved in the formation of disulfide bridges resulting in the expression of homodimeric (140kDa) surface receptors (Fig. 1). The 95 amino acid cytoplasmic tail, similar to all the other known inhibitory receptors, contains two classical ITIMs involved in the regulatory function of the p140 receptor.

Recently, a novel receptor belonging to the Ig superfamily has been identified by cDNA library screening (CANTONI et al. 1998b; SELVAKUMAR et al. 1996). This receptor, termed p49 (CANTONI et al. 1998b), displays at least 50% amino acid sequence identity with other NK receptors belonging to the same family and is encoded on chromosome 19 (SELVAKUMAR et al. 1996). It is characterized by two extracellular Ig-like domains of the C2 type and a 115 amino acid cytoplasmic tail containing a single ITIM (Fig. 1). This receptor may also exist in soluble form, since a cDNA has been isolated that is characterized by the absence of the exon encoding the transmembrane portion (CANTONI et al. 1998b). The specificity of the p49 receptor was determined by the generation of a chimerical protein formed by the ectodomain of p49 and the Fc portion of human IgG1 (p49-Fc). These soluble receptors bound efficiently to LCL721.221 cells transfected with HLA-G1, -A3, and -B46 alleles and weakly to the -B7 allele. On the other hand, they did not bind to the same cells when either untransfected or transfected with HLA-A2, -B51, -Cw3, or -Cw4. Since there was no correlation between the ability of p49 to recognize HLA class I alleles expressed on cell transfectants and the ability of those alleles to stabilize HLA-E expression (see below), these data suggest that the p49 receptor is able to bind HLA-G1 alleles directly and discriminate among allelic forms of classical HLA class I molecules (CANTONI et al. 1998b).

3 Inhibitory NK Receptors Belonging to the C type Lectin Superfamily

These receptors are type II transmembrane proteins containing a C type lectin domain. They are represented by heterodimers formed by the invariant CD94 molecule and different members of the NKG2 molecular family. The CD94 molecule was originally described as a disulfide-linked homodimer displaying a molecular mass of ~70kDa under nonreducing and 43kDa under reducing conditions (ARAMBURU et al. 1990). CD94 is expressed on most NK cells and on a minor subset of T cells. The first evidence indicating that this receptor could be involved in HLA recognition was provided by experiments showing that it could recognize HLA-Bw6 alleles but not HLA-Bw4 alleles (MORETTA et al. 1994). Indeed, target cells transfected with HLA-B7 (Bw6) were protected from lysis, whereas cells transfected with HLA-B*2705 (Bw4) were not. This inhibitory function was not detected in all CD94 + cells, suggesting that heterogeneity may have existed among the CD94 receptors expressed by different NK clones. In the same study, we could demonstrate that the inhibitory receptor for Bw6 was confined to a subset displaying the CD94^{bright} phenotype. Only NK clones displaying this phenotype were inhibited in redirected lysis against P815 targets in the presence of anti-CD94 mAbs (MORETTA et al. 1994). Subsequent studies indicated that clones that were not inhibited (CD94^{dull}) expressed an activating type of CD94 receptor (PERZ-VILLAR et al. 1995) characterized by a molecular mass of 39kDa instead of 43kDa (PERÉZ-VILLAR et al. 1996). Next, we generated mAbs able to distinguish between inhibitory and activating forms of the CD94 receptor. We showed that the Z199 and Z270 mAbs specifically react with the inhibitory receptor and that this receptor is confined to a defined NK subset (SIVORI et al. 1996). In the same study, it became clear that the specificity of the inhibitory CD94 receptor was not confined to certain Bw6 alleles and that HLA-A and HLA-C alleles were also able to protect target cells from lysis mediated by Z199 + /Z270 + NK cell clones. These studies, confirmed also by others (LAZETIC et al. 1996), suggested that this receptor could be characterized by a broad specificity for most HLA class I alleles (except those of BW4 supertypic specificity). The cDNA encoding CD94 (CHANG et al. 1995) revealed the existence of a type II transmembrane protein encoded by a single copy gene of the C type lectin superfamily located on human chromosome 12. Importantly, the short cytoplasmic tail of the cloned CD94 molecule did not contain consensus sequences for ITIM or ITAM (Fig. 1). Moreover, the cloned CD94 molecule was distinct from both the 43kDa and 39kDa proteins, since it resulted in a chain displaying a molecular mass of approximately 30kDa (PHILLIPS et al. 1996) that was poorly labeled by surface iodination. More recently, it became clear that the CD94 molecule could associate in heterodimers with either the 43-kDa or the 39-kDa molecules. The 43-kDa glycoprotein is the product of the NKG2 A gene (CARRETERO et al. 1997; LAZETIC et al. 1996; Fig. 1), whereas the 39-kDa glycoprotein is encoded by the NKG2 C gene (CANTONI et al. 1998a). The NKG2 genes (HOUCHINS et al. 1991) belong to a multigene family located on chromosome 12 in the so-called "human NK gene complex" (the genes of CD94, CD69, and human NKR-P1 are also located in this complex). The NKG2 A and NKG2 C glycoproteins are type II transmembrane receptors that display a C type lectin domain. They differ essentially in their cytoplasmic tail, since NKG2 A contains two ITIM sequences, whereas NKG2 C does not. We demonstrated that the NKG2 A glycoprotein is specifically recognized by the mAbs Z199 or Z270 directed against the inhibitory form of receptors, whereas the NKG2 C protein could only be identified by a new mAb (P25) that reacts with both NKG2 A and NKG2 C receptors (CANTONI et al. 1998a). Regarding the specificity of the CD94/NKG2 A receptor for HLA class I molecules, recent studies have indicated that this heterodimeric receptor is able to recognize HLA class Ib molecules rather than classical HLA class I molecules (BRAUD et al. 1997; BRAUD et al. 1998; LEE et al. 1998; BORREGO et al. 1998). In particular, it appears that this receptor recognizes HLA-E molecules both in soluble form and when expressed at the cell surface. It has been demonstrated that HLA-E expression on the cell membrane can take place only in the presence of a particular set of peptides. These are derived from the leader sequence of various HLA class I molecules that stabilize the conformation of HLA-E as a trimeric complex at the cell surface. Interestingly, the HLA class I-derived leader sequences able to stabilize HLA-E are those found in all classical HLA class I-molecules, with the exception of those derived from HLA-B alleles belonging to the Bw4 supertype. Consequently, HLA-E expression will be possible only when classical HLA class I molecules (except HLA-Bw4) are coexpressed at the cell surface. These findings, while demonstrating the fine specificity of the CD94/NKG2 receptors, also provide an explanation for previous data indicating that different HLA class I alleles (except HLA-Bw4) were capable of protecting target cells from NK-mediated cytotoxicity by CD94/NKG2A+ NK cells.

4 Triggering NK Receptors

As mentioned above, NK cells have evolved a number of receptor structures able to exert inhibitory control on their cytolytic function. The importance of these inhibitory mechanisms is well-demonstrated by the effect of masking the inhibitory receptor. Under these experimental conditions, it is likely that NK cells cannot be inhibited by negative signals generated by the interaction between inhibitory receptors and the corresponding protective HLA class I alleles. In the absence of such inhibitory signals, NK cells are allowed to kill target cells, suggesting the existence of receptors that, upon recognition of specific ligands on target cells, can trigger the cytolytic machinery. What are the nature and specificity of these receptors?

4.1 Triggering NK Receptors for Non-HLA Ligands

The fact that NK cells can also kill HLA-negative targets indicates that at least some of the receptors involved in NK cell triggering must operate in a non-MHC specific contest. The activity of these triggering receptors is normally under the negative control of inhibitory receptors specific for HLA class I; however, their functions become evident when recognition of the ligands for the inhibitory receptor is affected. Experimentally, this can be achieved by masking the inhibitory receptors with specific mAbs (see above). Physiologically, this could occur in tumor or virally transformed cells that undergo downregulation of the surface expression of one or more HLA class I alleles. So far, the nature and specificity of these non-MHC specific receptors have remained elusive. Several surface molecules that can mediate NK cell triggering have been identified. However, their actual role in natural cytotoxicity has still to be clarified, since, in most instances, these activating structures are not NK restricted. Among these triggering structures, the CD2 (BOLHUIS et al. 1986; LANIER et al. 1997), the CD69 (MORETTA et al. 1991), the p70 molecule recognized by the pp35 mAb (MORETTA et al. 1992), and DNAM-1 (SHIBUYA et al. 1996) are expressed on NK cells, but also on variable proportions of T lymphocytes. The low affinity receptor for the Fc portion of IgG (FcγRIII or CD16) is expressed on most NK cells and on a small subset of T lymphocytes (LANIER et al. 1996). CD16 is a potent inducer of NK cell triggering (MORETTA et al. 1989; LANIER et al. 1998); however, there is no evidence so far that it has functions other than ADCC against IgG-coated target cells. In this context, although inhibition of natural cytotoxicity mediated by human NK clones has been demonstrated by the use of anti-CD16 mAbs of the IgM isotype (MORETTA et al. 1989), it is not clear whether this reflects the masking of a putative triggering receptor or the induction of apoptosis in the effector cell population (ORTALDO et al. 1995; JEWETT et al. 1997).

Recently we have focused our studies on novel putative NK receptor structures characterized by the ability to trigger NK-mediated cytotoxicity and by a cell distribution highly restricted to NK cells. The search for these receptor structures led us to identification of two novel triggering receptors termed NKp46 (SIVORI et al. 1997) and NKp44 (VITALE et al. 1998). The NKp46 receptor is strictly confined to both resting and activated NK cells. This receptor could be identified by the generation of a specific mAb (BAB281) that was isolated based on the ability to trigger NK cell cytotoxicity.

Indeed, upon mAb-mediated cross-linking, NKp46 molecules not only trigger cytotoxicity but also induce $[Ca^{2+}]_i$ increases and lymphokine production. The NKp46-mediated NK cell activation is downregulated (and is controlled under physiological conditions) by the simultaneous engagement of inhibitory HLA Class-I-specific NK receptors (SIVORI et al. 1997). Molecular cloning of the NKp46 receptor by an expression cloning strategy allowed us to isolate a cDNA encoding a type I transmembrane glycoprotein belonging to the Ig superfamily (PESSINO et al. 1998) (Fig. 2). This receptor is encoded on human chromosome 19 and characterized, in the extracellular portion, by two Ig-like domains of the C2-type, in the transmembrane region by the presence of a charged aminoacid (Arg), and in the short cytoplasmic tail by the absence of ITAMs. Thus, NKp46 requires the association with ITAM-containing molecules that allow the transduction of signaling events that result in triggering NK-mediated cytotoxicity. The ITAM-containing subunits associated with NKp46 are represented by CD3ζ and FcεRIγ (VITALE et al. 1998). The NKp46-associated CD3ζ chains become tyrosine phosphorylated upon treatment with Pervanadate. Importantly, analysis of NKp46 expression by RT-PCR confirmed that this receptor is strictly confined to NK cells (PESSINO et al. 1998). The unique cellular distribution and the function of the NKp46 receptor suggest that it may be a candidate for a triggering receptor involved in the rec-

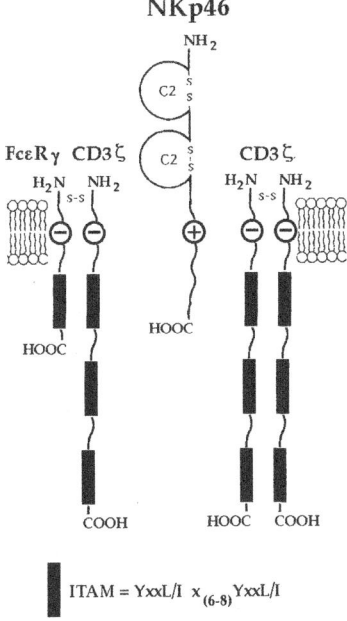

Fig. 2. Molecular structure of the NKp46 receptor

ognition of non-MHC ligands expressed on NK-susceptible targets. In this context, a possible view of the mechanisms of NK cell triggering against HLA-negative target cells is that multiple receptors may cooperate to induce optimal NK cell triggering (MORETTA et al. 1997b; LANIER et al. 1997). This does not imply that all triggering receptors function simultaneously, since their function may depend upon the presence and/or the density of their specific ligands on target cells. In line with this hypothesis, NKp46 masking by the specific mAb could significantly inhibit the cytolytic activity of NK cell clones against NK-susceptible target cells. However, in some cases, the inhibition was only partial, depending upon the target cells analyzed (SIVORI et al. 1997; PESSINO et al. 1998). This suggests that NKp46 may not be the unique activating receptor, but that it may cooperate with other triggering receptors to induce optimal NK cell activation (MORETTA 1997a). Along this line, the identification of another triggering receptor termed NKp44 (VITALE et al. 1998), allowed verification of this hypothesis. Thus, at least in some cases, the simultaneous masking of both NKp46 and NKp44 results in a cumulative inhibitory effect on non-MHC-restricted cytotoxicity. The NKp44 receptor, unlike NKp46, is selectively expressed by activated NK cells while it is lacking on resting NK cells. In addition, signal transduction via NKp44 does not require association with CD3ζ or FcεRIγ. On the other hand, the recently characterized KARAP/DAP12 subunits (OLCESE et al. 1997; LANIER et al. 1998b) were shown to associate with the NKp44 receptor (VITALE et al. 1998).

Notably, the analysis of several NK cell clones provides evidence of the existence of additional triggering receptors distinct from NKp46 and NKp44 (VITALE et al. 1998; PESSINO et al. 1998). Identification of such additional receptors will certainly help dissect the pathways of NK cell triggering and their respective role in the process of killing different NK-susceptible target cells.

4.2 Triggering NK Receptors for HLA Molecules

The surface receptors that can trigger NK-mediated cytotoxicity are not confined to those displaying non-HLA restricted specificity. In this context, tumor or virus transformed cells may undergo only a partial downregulation of HLA expression that may involve some, but not all HLA class I alleles. Under these conditions, the "protective" alleles may be lost while other "nonprotective" alleles may still be expressed at the cell surface. The latter molecules may represent target structures recognized by triggering receptors specific for HLA class I ligands. The first family of such receptors has been identified by the use of the same mAbs that are specific for HLA-C-specific inhibitory receptors. Thus, anti-p58.1 (CD158a) and anti-p58.2 (CD158b) mAbs were also found to react with surface molecules involved in NK cell activation. These molecules are characterized by a lower molecular mass (50kDa) and thus were termed p50.1 and p50.2 on the basis of their reactivity with EB6 (CD158a) or GL183 (CD158b) mAbs (MORETTA et al. 1995). Molecular cloning revealed, in the extracellular region, a high degree of sequence identity with the corresponding inhibitory (p58) receptors. However, while the inhibitory forms are characterized by a nonpolar transmembrane region, the p50 activating receptors are characterized by the charged amino acid Lys (BIASSONI et al. 1996) (Fig. 3). The cytoplasmic tail is shorter (39 amino acids) and does not contain ITIM sequences. The activating p50 receptors were shown to display the same HLA-C specificity as their inhibitory counterparts (MORETTA et al. 1995). A third member of this family, termed p50.3, has also been characterized. This activating receptor is recognized by specific mAbs termed PAX180 and FES172. In contrast to p50.1 and p50.2, no inhibitory counterpart of this receptor has been identified (BOTTINO et al. 1996). Similar to all the other activating NK receptors, the function of p50 molecules is also regulated, i.e., suppressed, by inhibitory receptors which are coexpressed with p50 at the NK cell surface. (MORETTA et al. 1995; BOTTINO et al. 1996). Thus, it is likely that, under physiological conditions, the activating function of p50 receptors can take place only when the coexpressed inhibitory receptor(s) cannot bind to the protective alleles. The fact that inhibitory receptors predominate over the triggering ones may be important in preventing the occurrence of autoimmune phenomena against normal autologous cells.

The precise physiological role of these activating receptors is still undefined, although in vivo situations of selective downregulation of protective alleles (GARRIDO et al. 1997) may favor the triggering of p50 molecules by the remaining, nonprotective alleles. Since p50 and p58 recognize the same HLA-C ligands, it is also possible that they may sense different HLA-C-bound peptides and that

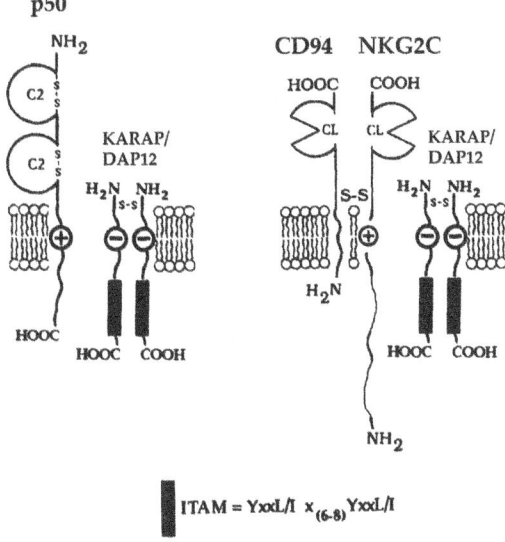

Fig. 3. HLA-specific triggering NK receptors

different peptides may influence the affinity of the receptor for its ligand. In this context, we showed that soluble p50.1 and p58.1 receptor molecules bind to Cw4 molecules, although with different affinity (BIASSONI et al. 1997). Moreover, single amino acid substitution may be responsible for their differential affinity. We also demonstrated that activating p50 receptors are included in a multimeric complex with a new family of phosphorylated proteins that have been termed killer activating receptor-associated proteins (KARAPS) (OLCESE et al. 1997). These proteins become tyrosine phosphorylated upon receptor cross-linking and have a molecular mass of approximately 12–16kDa. Recently, molecular cloning revealed that KARAPS (DAP12) molecules are encoded by a gene localized on human chromosome 19 and characterized by having transmembrane regions with a charged residue (Asp) and a cytoplasmic tail containing an ITAM sequence (LANIER et al, 1998a). Their function is likely to be crucial for NK cell triggering via the p50 receptors. Indeed, experiments with reconstitution of p50 expression in the absence of KARAPS indicated that p50 molecules alone are unable to transduce any detectable triggering signal (BLERY et al. 1997).

By using the newly identified P25 mAb, it recently became possible to characterize the activating form of the CD94/NKG2 heterodimer (CANTONI et al. 1998a). In addition to NKG2 A (inhibitory chain), the P25 mAb also recognize NKG2 C molecules. This mAb, in NK clones characterized by an activating form of the CD94/NKG2 receptors, identifies a CD94-associated molecule of approximately 39kDa. The surface expression of NKG2 C, similarly to NKG2 A, was dependent on CD94 expression, since cell transfection with NKG2 C alone did not result in surface expression of P25-reactive molecules. NKG2 A and NKG2 C molecules display a high degree of sequence identity in extracellular regions, while the NKG2 C transmembrane and cytoplasmic portions are characterized, respectively, by a charged residue (Lys) and the lack of ITIM sequences (Fig. 3). This is reminiscent of the differences found between p58 and p50 receptors (BIASSONI et al. 1996). As with p50 and NKp44, signaling via the CD94/NKG2 C receptor is dependent upon the association to KARAP/DAP12 ITAM-bearing molecules (LANIER et al. 1998b). Based on the high degree of homology between the ecto-domains of NKG2 A and NKG2 C, it is conceivable that the specificity of the CD94/NKG2 C receptor may be identical to that previously demonstrated for CD94/NKG2 A receptors (BRAUD et al. 1998). However, functional analysis of NK clones expressing CD94/NKG2 C indicates that this receptor is preferentially responsive towards certain HLA class I transfectants (i.e., HLA-G1 +), while it is moderately responsive against other cell transfectants expressing comparable amounts of HLA-E-stabilizing HLA class I molecules, i.e., HLA-B7 + and HLA-A2 + (Moretta, unpublished observation).

It is notable that the P25 mAb, although reactive with both NKG2 A and NKG2 C, does not bind to all CD94 + NK cells. These P25-/CD94 + cells are characterized by a CD94 receptor displaying either low levels of triggering function or no function at all. In the first case, the CD94 molecule might be associated with another member of the NKG2 family that does not react with the p25 mAb, such as NKG2 E. In the second case, it is possible that the receptor may be formed by homodimers of CD94. In both types of NK clones, however, immunoprecipitation with anti-CD94 mAbs did not reveal CD94-associated chains. This may be at least in part a result of the low levels of CD94 expression that usually characterize these P25-/CD94 + NK clones.

References

Aramburu J, Balboa MA, Ramirez A, Silva A, Acevedo A, Sànchez-Madrid F, De Landàzuri MO, Lòpez-Botet M (1990) A novel functional cell surface dimer (Kp43) expressed by natural killer cells and T cell receptor-γ/δ + T lymphocytes. Inhibition of IL-2 dependent proliferation by anti-Kp43 monoclonal antibody. J Immunol 44:3238

Biassoni R, Falco M, Cambiaggi A, Costa P, Verdiani S, Pende D, Conte R, Di Donato C, Parham P, Moretta L (1995) amino acid substitutions can influence the natural killer (NK)-mediated recognition of HLA-C molecules. Role of serine-77 and lysine-80 in the target cell protection from lysis mediated by "Group 2" or "Group 1" NK clones. J Exp Med 182:605

Biassoni R, Cantoni C, Falco M, Verdiani S, Bottino C, Vitale M, Conte R, Poggi A, Moretta A, Moretta L (1996) The human leukocyte antigen (HLA)-C-specific "activatory" or "inhibitory" natural killer cell receptors display highly homologous extracellular domains but differ in their transmembrane and intracytoplasmic portions. J Exp Med 183:645

Biassoni R, Pessino A, Malaspina A, Cantoni C, Sivori S, Moretta L, Moretta A (1997) Role of amino acid position 70 in the binding affinity of p50.1 and p58.1 receptors for HLA-Cw4 molecules. Eur J Immunol pp3095–3099

Binstadt BA, Brumbaugh KM, Dick CJ, Scharenberg AM, Williams BL, Colonna M, Lanier LL, Kinet JP, Abraham RT, Leibson PJ (1996) Sequential involvement of Lck and SHP-1 with MHC-recognizing receptors on NK cells inhibits FcR-initiated tyrosine kinase activation. Immunity 5:629

Bléry M, Delon J, Trautman A, Cambiaggi A, Olcese L, Biassoni R, Moretta L, Moretta A, Däeron M, Vivier E (1997) Reconstituted killer-cell-inhibitory receptors for MHC class I molecules control mast cell activation induced via immunoreceptor tyrosine-based activation motifs. J Biol Chem 272:8989–8996

Bolhuis RLH, Roozemond RC, Van de Griend RJ (1986) Induction and blocking of cytolysis in CD2+, CD3- NK and CD2+, CD3+ cytotoxic T lymphocytes via the CD2 50kD sheep erythrocyte receptor. J Immunol 136:3939

Borrego F, Ulbrecht M, Weiss EH, Coligan JE, Brooks, AG (1998) Recognition of human histocompatibility leukocyte antigen (HLA)-E complexed with HLA class I signal sequence-derived peptides by CD94/NKG2 confers protection from natural-killer-cell-mediated lysis. J Exp Med 187:813–818

Bottino C, Vitale M, Olcese L, Sivori S, Morelli L, Augugliaro R, Ciccone E, Moretta L, Moretta A (1994) The human natural killer cell receptor for major histocompatibility complex class I molecules. Surface modulation of p58 molecules and their linkage to CD3z chain, FcεRIγ chain, and the p56^{lck} kinase. Eur J Immunol 24:2527

Bottino C, Sivori S, Vitale M, Cantoni C, Falco M, Pende D, Morelli L, Augugliaro R, Semenzato GP, Biassoni R, Moretta L, Moretta A (1996) A novel surface molecule homologous to the p58/p50 family of receptors is selectively expressed on a subset of human natural killer cells and induces both triggering of cell functions and proliferation. Eur J Immunol 26:1816

Braud VM, Yvonne Jones E, McMichael A (1997) The human major histocompatibility complex class Ib molecule HLA-E binds signal-sequence-derived peptides with primary anchor residues at positions 2 and 9. Eur J Immunol 27:1164–1169

Braud VM, Allan DSJ, O'Callaghan CA, Soderstrom K, D'Andrea A, Ogg GS, Lazetic S, Young NT, Bell JI, Phillips JH, Lanier LL, McMichael AJ (1998) HLA-E binds to natural killer cell receptors CD94/NKG2 A, B, and C. Nature 391:795–799

Burshtyn DN, Sharenberg AM, Wagtmann N, Rajagopalan S, Berrula K, Yi T, Kinet J-P, Long EO (1996) Recruitment of tyrosine phosphatase HCP by the killer cell inhibitory receptor. Immunity 4:77

Campbell KS, Dessing M, Lopez-Botet M, Cella M, Colonna M (1996) Tyrosine phosphorylation of a human killer inhibitory receptor recruits protein phosphatase 1 C. J Exp Med 184:93

Cantoni C, Verdiani S, Falco M, Conte R, Biassoni R (1996) Molecular structures of HLA-specific human NK cell receptors. Chem Immunol 64:88

Cantoni C, Biassoni R, Pende D, Sivori S, Accame L, Moretta L, Moretta A, Bottino C (1998a) The activating form of CD94 receptor complex. The CD94-associated Kp39 protein represents the product of the NKG2-C gene. Eur J Immunol 28:327–338

Cantoni C, Verdiani S, Falco M, Pessino A, Cilli M, Conte R, Pende D, Ponte M, Mikaelsson MS, Moretta L, Biassoni R (1998b) p49, a novel putative HLA-class-I-specific inhibitory NK receptor belonging to the immunoglobulin superfamily. Eur J Immunol 28:1980–1990

Carretero M, Cantoni C, Bellón T, Bottino C, Biassoni R, Rodríguez A, Pérez-Villar JJ, Moretta L, Moretta A, López-Botet M (1997) The CD94 and NKG2 A C type lectins covalently assemble to form a NK cell inhibitory receptor for HLA class I molecules. Eur J Immunol 27:563

Chang C, Rodriguez A, Carretero M, Lòpez-Botet M, Phillips JH, Lanier LL (1995) Molecular characterization of human CD94: a type II membrane glycoprotein related to the C type lectin superfamily. Eur J Immunol 25:2433

Ciccone E, Pende D, Viale O, Di Donato C, Orengo AM, Biassoni R, Verdiani S, Amoroso A, Moretta A, Moretta L (1992a) Involvement of HLA class I alleles in NK-cell-specific function: expression of HLA-Cw3 confers selective protection from lysis by alloreactive NK clones displaying a defined specificity (specificity 2). J Exp Med 176:963–971

Ciccone E, Pende D, Viale O, Di Donato C, Tripodi G, Orengo AM, Guardiola J, Moretta A, Moretta L (1992b) Evidence of a natural killer (NK) cell repertoire for (allo)antigen recognition: definition of five distinct NK-determined allospecificities in humans. J Exp Med 175:709

Ciccone E, Pende D, Vitale M, Nanni L, Di Donato C, Bottino C, Morelli L, Viale O, Amoroso A, Moretta A, Moretta L (1994) Self class I molecules protect normal cells from lysis mediated by autologous natural killer cells. Eur J Immunol 24:1003

Colonna M, Samaridis J (1995) Cloning of immunoglobulin superfamily members associated with HLA-C and HLA-B recognition by human natural killer cells. Science 268:405

Colonna M, Borsellino G, Falco M, Ferrara GB, Strominger JL (1993) HLA-C is the inhibitory ligand that determines dominant resistance to lysis by NK1- and NK2-specific natural killer cells. Proc Natl Acad Sci USA 90:12000

D'Andrea A, Chang C, Franz-Bacon K, McClanahan T, Phillips JH, Lanier LL (1995) Molecular cloning of NKB1. A natural killer cell receptor for HLA-B allotypes. J Immunol 155:2306

Döhring C, Scheidegger D, Samaridis J, Cella M, Colonna M (1996) A human killer inhibitory receptor specific for HLA-A. J Immunol 156:3098

Fry AM, Lanier LL, Weiss A (1996) Phosphotyrosines in the killer cell inhibitory receptor motif of NKB1 are required for negative signalling and for association with protein tyrosine phosphatase 1 C. J Exp Med 184:295

Garrido F, Cabrera T, Lopez Nevot MA, Ruiz-Cabello F (1995) HLA-class I antigens in human tumours. Adv Cancer Res 67:155

Garrido F, Ruiz-Cabello F, Cabrera T, Pérez-Villar JJ, Lòpez-Botet M, Duggan-Keen M, Stern PL (1997) Implications for immunosurveillance of altered HLA class I phenotypes in human tumours. Immunol Today 18:89

Gumperz JE, Litwin V, Phillips JH, Lanier LL, Parham P (1995) The Bw4- public epitope of HLA-B molecules confers reactivity with natural killer cell clones that express NKB1, a putative HLA receptor. J Exp Med 181:1113

Herberman RB, Nunn ME, Lavrin DH (1975) Natural cytotoxic reactivity of mouse lymphoid cells against syngenic and allogeneic tumours. I: Distribution of reactivity and specificity. Int J Cancer 16:216

Houchins JP, Yabe T, McSherry C, Bach FH (1991) DNA sequence analysis of NKG2, a family of related cDNA clones encoding type II integral membrane proteins on human natural killer cells. J Exp Med 173:1017

Karre K (1992) An unexpected petition for pardon. Current Biology 11:613

Kiessling R, Klein E, Wigzell H (1975) "Natural killer" cells in the mouse I. Cytotoxic cells with specificity for mouse moloney leukemia cells. Specificity and distribution according to genotype. Eur J Immunol 5:112

Ikeda H, Lethé B, Lehmann F, Van Baren N, Baurain JF, De Smet C, Chambost H, Vitale M, Moretta A, Boon T, Coulie P (1997) Characterization of an antigen that is recognized on a melanoma showing partial HLA loss by CTL expressing an NK inhibitory receptor. Immunity 6:199

Jewett A, Cavalcanti M, Bonavida B (1997) Pivotal role of endogenous TNF-alpha in the induction of functional inactivation and apoptosis in NK cells. J Immunol 159:4815–4822

Lanier LL, Phillips J, Hackett J, Tutt M, Kumar V (1986) Natural killer cells: definition of a cell type rather than a function. J Immunol 137:2375

Lanier LL, Ruitenberg JJ, Phillips JH (1988) Functional and biochemical analysis of CD16 antigen on natural killer cells and granulocytes. J Immunol 141:3478

Lanier LL, Corliss B, Phillips JH (1997) Arousal and inhibition of human NK cells. Immunol Rev 155:145

Lanier LL, Corliss BC, Wu J, Leong C, Phillips JH (1998) Immunoreceptor DAP12 bearing a tyrosine-based activation motifs is involved in activating NK cells. Nature 391:703–707

Lanier LL, Corliss B, Wu J, Phillips JH (1998b) Association of DAP12 with the activating CD94/NKG2 C NK cell receptors. Immunity 8:693–701

Lazetic S, Chang C, Houchins JP, Lanier LL, Phillips JH (1996) Human natural killer cell receptors involved in MHC class I recognition are disulphide-linked heterodimers of CD94 and NKG2 subunits. J Immunol 157:4741

Lee N, Llano M, Carretero M, Ishitani A, Navarro F, Lòpez-Botet M, Gherarty D (1998) HLA-E is a major ligand for the natural killer inhibitory receptor CD94/NKG2 A. Proc Natl Acad Sci USA 95:5199–5204

Litwin V, Gumperz JE, Parham P, Phillips JH, Lanier LL (1994) NKB1: a natural killer cell receptor involved in the recognition of polymorphic HLA-B molecules. J Exp Med 180:537

Madden DR, Gorga JC, Strominger JL, Wiley DC (1992) The three-dimensional structure of HLA-B27 at 2.1 A resolution suggests a general mechanism for tight peptide binding to MHC. Cell 70:1035

Malnati MS, Peruzzi M, Parker KC, Biddison WE, Ciccone E, Moretta A, Long EO (1995) Peptide specificity in the recognition of MHC class I by natural killer cell clones. Science 267:1016

Moretta A (1997a) Molecular mechanisms in cell-mediated cytotoxicity. Cell 90:13

Moretta A, Tambussi G, Ciccone E, Pende D, Melioli G, Moretta L (1989) CD16 Surface molecules regulate the cytolytic function of CD3-CD16+ human "natural killer" cells. Int J Cancer 44:727

Moretta A, Bottino C, Pende D, Tripodi G, Tambussi G, Viale O, Orengo A, Barbaresi M, Merli A, Ciccone E, Moretta L (1990a) Identification of four subsets of human CD3-CD16+ NK cells by the expression of clonally distributed functional surface molecules. Correlation between subset assignment of NK clones and ability to mediate specific alloantigen recognition. J Exp Med 172:1589

Moretta A, Tambussi G, Bottino C, Tripodi G, Merli A, Ciccone E, Pantaleo G, Moretta L (1990b) A novel surface antigen expressed by a subset of human CD3-CD16+ natural killer cells. Role in cell activation and regulation of cytolytic function. J Exp Med 171:695

Moretta A, Poggi A, Pende D, Tripodi G, Orengo AM, Pella N, Augugliaro R, Bottino C, Ciccone E, Moretta L (1991) CD69-mediated pathway of lymphocyte activation. Anti-CD69 mAbs trigger the cytolytic activity of different lymphoid effector cells, with the exception of cytolytic T lymphocytes expressing TCRα/β. J Exp Med 174:1393

Moretta L, Ciccone E, Moretta A, Höglund P, Ohlen C, Karre K (1992) Allorecognition by NK cells: nonself or no self? Immunology Today 13:300

Moretta A, Bottino C, Tripodi G, Vitale M, Pende D, Morelli L, Augugliaro R, Barbaresi M, Ciccone E, Millo R, Moretta L (1992) Novel surface molecules involved in human NK cell activation and triggering of the lytic machinery. Int J Cancer Suppl 7:6

Moretta A, Vitale M, Bottino C, Orengo AM, Morelli L, Augugliaro R, Barbaresi M, Ciccone E, Moretta L (1993) P58 molecules as putative receptors for MHC class I molecules in human natural killer (NK) cells. Anti-p58 antibodies reconstitute lysis of MHC class-I-protected cells in NK clones displaying different specificities. J Exp Med 178:597

Moretta L, Ciccone E, Mingari MC, Biassoni R, Moretta A (1994). Human NK cells: origin, clonality, specificity, and receptors. Adv Immunol vol 55:341

Moretta A, Vitale M, Sivori S, Bottino C, Morelli L, Augugliaro R, Barbaresi M, Pende D, Ciccone E, Lopez-Botet M, Moretta L (1994) Human natural killer cell receptors for HLA-class I molecules. Evidence that the Kp43 (CD94) molecule functions as receptor for HLA-B alleles. J Exp Med 180:545

Moretta A, Sivori S, Vitale M, Pende D, Morelli L, Augugliaro R, Bottino C, Moretta L (1995) Existence of both inhibitory (p58) and activatory (p50) receptors for HLA-C molecules in human natural killer cells. J Exp Med 182:875

Moretta A, Bottino C, Vitale M, Pende D, Biassoni R, Mingari MC, Moretta L (1996) Receptors for HLA class-I-molecules in human natural killer cells. Ann Rev Immunol 14:619

Moretta A, Biassoni R, Bottino C, Pende D, Vitale M, Poggi A, Mingari MC, Moretta L (1997b) Major histocompatibility complex class-I-specific receptors on human natural killer and T lymphocytes. Immunol Rev 155:105–117

Moretta A, Bottino C, Biassoni R (1997c) Natural killer receptors for HLA-C alleles. Proceedings of the 6th International Workshop and Conference, Garland publishing, pp 290–292

Olcese L, Lang P, Vély F, Cambiaggi A, Marguet D, Bléry M, Hippen KL, Biassoni R, Moretta A, Moretta L, Cambieri JC, Vivier E (1996) Human and mouse natural killer cell inhibitory receptors recruit the PTP1C and PTP1D protein tyrosine phosphatases. J Immunol 156:4531

Olcese L, Cambiaggi A, Semenzato G, Bottino C, Moretta A, Vivier E (1997) Human killer cell activatory receptors for MHC class I molecules are included in a multimeric complex expressed by natural killer cells. J Immunol 158:5083

Ortaldo JR, Mason AT, O'Shea JJ (1995) Receptor-induced death in human natural killer cells: involvement of CD16. J Exp Med 181:339–344

Pende D, Biassoni R, Cantoni C, Verdiani S, Falco M, Di Donato C, Accame L, Bottino C, Moretta A, Moretta L (1996) The natural killer cell receptor specific for HLA-A allotypes: a novel member of the p58/p70 family of inhibitory receptors that is characterized by three immunoglobulin-like domains and is expressed as a 140kD disulphide-linked dimer. J Exp Med Vol 184:505

Perez-Villar JJ, Melero I, Rodríguez A, Carretero M, Aramburu J, Sivori S, Orengo AM, Moretta A, Lòpez-Botet M (1995) Functional ambivalence of the Kp43 (CD94) NK-cell-associated surface antigen. J Immunol 154:5779

Perez-Villar JJ, Carretero M, Navarro F, Melero I, Rodriguez A, Bottino C, Moretta A, Lòpez-Botet M (1996) Biochemical and serological evidence for the existence of functionally distinct isoforms of the CD94 NK-cell receptor. J Immunol 157:5367

Peruzzi M, Parker KC, Long EO, Malnati MS (1996) Peptide sequence requirements for the recognition of HLA-B*2705 by specific natural killer cells. J Immunol 157:3350

Pessino A, Sivori S, Bottino C, Malaspina A, Morelli L, Moretta L, Biassoni R, Moretta A (1998) Molecular cloning of NKp46: a novel member of the immunoglobulin superfamily involved in triggering of natural cytotoxicity. J Exp Med (in press)

Phillips JH, Chang C, Mattson J, Gumperz JE, Parham P, Lanier LL (1996) CD94 and a novel associated protein (94AP) form a NK cell receptor involved in the recognition of HLA-A, -B, and -C allotypes. Immunity 5:163

Rajagopalan S, Long EO (1997) The direct binding of a p58 killer cell inhibitory receptor to human histocompatibility leukocyte antigen (HLA)-Cw4 exhibits peptide selectively. J Exp Med 185:1523

Selvakumar S, Steffens U, Dupont B (1996) NK cell receptor gene of the KIR family with two IG domains but highest homology to KIR receptors with three IG domains. Tissue Antigens 48:285–295

Shibuya A, Campbell D, Hannum C, Ylssel H, Franz-Bacon K, McClanahan T, Kitamura T, Nicholl J, Sutherland GR, Lanier LL, Phillips JH (1996) DNAM-1, a novel adhesion molecule involved in the cytolytic function of T lymphocytes. Immunity 4:573–581

Sivori S, Vitale M, Bottino C, Marcenaro E, Sanseverino L, Parolini S, Moretta L, Moretta A (1996) CD94 functions as a natural killer cell inhibitory receptor for different HLA-CLASS-I alleles. Identification of the inhibitory form of CD94 by the use of novel monoclonal antibodies. Eur J Immunol 26:2487

Sivori S, Vitale M, Morelli L, Sanseverino L, Augugliaro R, Bottino C, Moretta L, Moretta A (1997) p46, a novel natural-killer-cell-specific surface molecule that mediates cell activation. J Exp Med 186: 1129–1136

Trinchieri G (1989) Biology of natural killer cells. Adv Immunol 47:187

Valiante NM, Phillips JH, Lanier LL, Parham P (1996) Killer cell inhibitory receptor recognition of human leukocyte antigen (HLA) class I blocks formation of a pp36/PLC-g signaling complex in human natural killer (NK) cells. J Exp Med 184:2243–2250

Valiante NM, Lienert K, Shilling HG, Smits BJ, Parham P (1997) Killer cell receptors: keeping pace with MHC class I evolution. Immunol Rev 155:155

Vély F, Olivero S, Olcese L, Moretta A, Damen JE, Liu L, Krystal G, Cambier JC, Daëron M, Vivier E (1997) Differential association of phosphatases with hematopoietic coreceptors bearing immunoreceptor tyrosine-based inhibition motivs. Eur J Immunol 27:1994

Vitale M, Sivori S, Pende D, Moretta L, Moretta A (1995) The coexpression of two functionally independent p58 inhibitory receptors in human NK cell clones results in the inability to kill all normal allogeneic target cells. Proc Natl Acad Sci USA 92:3536

Vitale M, Sivori S, Pende D, Augugliaro R, Di Donato C, Amoroso A, Malnati M, Bottino C, Moretta L, Moretta A (1996) Physical and functional independency of p70 and p58 cell receptors for HLA-class I. Their role in the definition of different groups of alloreactive NK cell clones. Proc Natl Acad Sci USA 93:1453

Vitale M, Bottino C, Sivori S, Sanseverino L, Castriconi R, Marcenaro R, Augugliaro R, Moretta L, Moretta A (1998) NKp44, a novel triggering surface molecule specifically expressed by activated natural killer cells, is involved in non-MHC-restricted tumor cell lysis. J Exp Med 187:2065–2072

Wagtmann N, Biassoni R, Cantoni C, Verdiani S, Malnati MS, Vitale M, Bottino C, Moretta L, Moretta A, Long EO (1995) Molecular clones of the p58 NK cell receptor reveal immunoglobulin-related molecules with diversity in both the extra- and intracellular domains. Immunity 2:439

Weber JR, Orstavik S, Torgersen KM, Danbolt NC, Berg JF, Tasken K, Imboden JB, Vaage JT (1998) Molecular cloning of the cDNA-encoding pp36, a tyrosine phosphorylated adaptor protein selectively expressed by T cells and natural killer cells. J Exp Med 187:1157–1161

Zappacosta F, Borrego F, Brooks AG, Parker KC, Coligan JE (1997) Peptides isolated from HLA-Cw*0304 confer different degrees of protection from natural-killer-cell-mediated lysis. Proc Natl Acad Sci USA 94:6313

Regulation of NK Cell Functions Through Interaction of the CD94/NKG2 Receptors with the Nonclassical Class I Molecule HLA-E

V.M. Braud and A.J. McMichael

1 Introduction

Natural killer (NK) cells constitute 5%–20% of the lymphocyte population in the periphery. They are cytotoxic, but also secrete cytokines. They play a major role in innate immunity, as an early defense against a variety of infections and influence the development of acquired immunity. They also participate in the elimination of certain tumors and virally infected cells which have lost or downregulated their MHC class I molecules. This concept was originally proposed by Kärre as "the missing self hypothesis" (Kärre 1997). The cytotoxic activity of NK cells is regulated by cell surface receptors for MHC class I molecules that can deliver activatory or inhibitory signals. The characterization of these receptors has considerably increased in the last few years (Lanier 1998). They are generally classified into two main families: the immunoglobulin superfamily and the C-type lectin superfamily. In humans, NK receptors belonging to the Ig superfamily can recognize groups of HLA class I molecules, such as two distinct groups of HLA-C alleles (KIR2DL1, KIR2DL2, KIR2DL3, KIR2DS1, KIR2DS2), one group of B alleles displaying Bw4 supertypic specificity (KIR3DL1), and the HLA-A alleles, HLA-A3 and HLA-A11 (KIR3DL2). CD94/NKG2 heterodimers belong to the C-type lectin superfamily and were thought to recognize most HLA class I molecules.

Institute of Molecular Medicine, Headington, Oxford OX3 9DS, UK

Activation of NK cytotoxicity is mediated by a variety of cell surface receptors and adhesion molecules, but the mechanisms for that are still unclear (BURSHTYN and LONG 1997). It appears, however, that activatory receptors which are specific for class I molecules, and which lack the immunoreceptor tyrosine-based activatory motif (ITAM), associate with a protein containing an ITAM in its cytoplasmic domain (OLCESE et al. 1997; LANIER et al. 1998a; LANIER et al. 1998b). Upon ligation, the ITAM sequence is tyrosine-phosphorylated by src-family tyrosine kinases and recruits ZAP-70 or Syk protein tyrosine kinases. This leads to a cascade of events triggering cytotoxicity or cytokine production. NK cellular activation is also downregulated by inhibitory signals. Upon engagement of the inhibitory receptors, the immunoreceptor tyrosine-based inhibitory motif (ITIM) in their cytoplasmic domain is phosphorylated, and recruits SH2 domain-containing protein tyrosine phosphatases (SHP) involved in the inhibition of cytotoxic activity.

2 The CD94/NKG2 Receptor Complex

The CD94/NKG2 receptor complexes are disulfide-linked heterodimers formed by the association of the invariant CD94 glycoprotein with a member of the NKG2 family (LAZETIC et al. 1996; BROOKS et al. 1997; CARRETERO et al. 1997; CANTONI et al. 1998). Four genes encode for the NKG2 glycoproteins: NKG2 A, NKG2 C, NKG2 E and NKG2 D/F genes. It has been shown that CD94 is essential to permit transport and expression of NKG2 glycoproteins onto the cell surface (LAZETIC et al. 1996; LANIER et al. 1998a). CD94 has a very short cytoplasmic domain and therefore cannot transduce signals. When CD94 is associated with NKG2 A (which contains ITIMs), the heterodimer forms an inhibitory receptor, whereas when associated with NKG2 C it forms a stimulatory receptor (HOUCHINS et al. 1997). A recent study showed that the CD94/NKG2 C activatory receptor also contains a third subunit, DAP12 (LANIER et al. 1998b; LANIER et al. 1998a). DAP12 is expressed as a disulfide-bonded homodimer and interacts with NKG2 C via charged residues in their transmembrane domains. DAP12 is necessary for transport of the CD94/NKG2 C complex to the cell surface. Ligation of CD94 on CD94/NKG2 C/DAP12 transfectants caused tyrosine phosphorylation of DAP12, suggesting that it induced cellular activation via DAP12, which displays an ITAM motif in its cytoplasmic domain.

A number of functional studies aimed to characterize the specificity of the inhibitory CD94/NKG2 A receptor. Using class I-deficient cell lines transfected with single HLA molecules, it appeared that CD94/NKG2 A could recognize most but not all classical class I molecules as well as the nonclassical HLA-G molecule (SODERSTROM et al. 1997; PEREZ VILLAR et al. 1997; SIVORI et al. 1996; PHILLIPS et al. 1996; PENDE et al. 1997). The specificity was puzzling at first, because the molecules recognized did not appear to share any amino acid sequence motif that was different from those not recognized.

Recent studies on the nonclassical class I molecule HLA-E have provided an explanation for the unclear specificity of CD94/NKG2 A receptors, in that HLA-E is the ligand for CD94/NKG2 A and CD94/NKG2 C receptors.

3 Identification of Peptides Bound to HLA-E

HLA-E is ubiquitously expressed and exhibits very limited polymorphism. HLA-E is highly homologous to classical class I molecules, but displays some unique substitutions in the peptide binding groove which are shared with another non-classical class I molecule, Qa-1 in mouse. Clues to the identification of peptides binding to HLA-E were provided by Qa-1. A peptide termed Qdm, derived from the leader sequence of H-2 D and H-2 L class I molecules (residues 3–11, AMAPRTLLL), was eluted from Qa-1 molecules and appeared to be the dominant peptide present (ALDRICH et al. 1994). Leader sequences in general are variable, but those of MHC class I molecules are highly conserved and a similar peptide (residues 3–11) can be generated from HLA-G and human classical MHC class I molecules. Such a peptide is not present in HLA-E and HLA-F molecules because they have shorter leader sequences. These HLA leader peptides were tested in a peptide-binding assay in vitro in which the ability of peptides to stabilize HLA-E molecules in cell lysates was measured (BRAUD et al. 1997). The mutant cell line 721.221, which only endogenously expresses HLA-E and HLA-F, was used. Whereas Qdm peptide was capable of stabilizing some HLA-E molecules when added at high concentration, most of the human signal-derived peptides bound with high affinity. However, the substitution of the methionine for a threonine at position 2 disrupted peptide binding. This amino acid is found in the leader sequence of two thirds of HLA-B alleles. The optimum peptide binding to HLA-E was the nonamer, although eight, ten, or eleven-mers could bind with lower affinity (BRAUD et al. 1997). Anchor residues were localized by substituting each amino acid residue with an alanine or a glycine. Positions 2 and 9 were found to be the primary anchor residues, whereas position 7 was a secondary anchor residue. Substitution of position 9, 10, or 11 in the 10- and 11-mer peptides revealed that the leucine at position 9 was still the anchor residue, suggesting that residues 10 and 11 were probably lying outside the binding groove. Interestingly, the substitution of each residue caused a significant reduction in the efficiency of peptide binding to HLA-E, suggesting that the peptide was binding throughout its entire length. This obser-vation was confirmed by the crystal structure of HLA-E, which was determined in a complex with a nonamer peptide (VMAPRTVLL) derived from the conserved residues 3–11 of certain classical class I molecules (O'CALLAGHAN et al. 1998). The binding groove retained well-defined pockets, but the peptide selectivity was en-hanced by occupancy of all of the pockets. One of the characteristics that HLA-E has in common with Qa-1, is the substitution of the highly conserved tryptophan 147 with a serine, eliminating the hydrogen bond from the side-chain nitrogen to the main-chain carbonyl oxygen of position 8 (P8) of the peptide. In HLA-E, this hydrogen bond is replaced by a more extensive network, pinning the peptide main chain into the groove. The side chain of Asn-77 plays a critical role, forming two hydrogen bonds to the peptide main chain (P7 and P9). The positioning of P6 and P7 into their respective pockets is also necessary to ensure stabilization of the peptide in a much lower position than peptides in classical class I binding grooves. The proline at position 4 contributes to the orientation of the peptide deep into the groove. Overall, the structure of the HLA-E binding groove imposes constraints on bound peptides. This observation correlates with the data obtained from the elution of peptides from a chimeric molecule consisting of the leader sequence of HLA-A2 and the extracellular, transmembrane, and cytoplasmic domains of HLA-E (LEE et al. 1998a). Interestingly, the leader sequence peptide from HLA-A2 signal sequence constituted 62% of the eluted material, suggesting that HLA-E was binding a restricted pool of peptides.

4 Regulation of HLA-E Expression

For many years, the study of HLA-E expression had been difficult, due to a lack of HLA-E specific antibodies. One study demonstrated that HLA-E was poorly expressed at the cell surface of the mouse cell line P3X63-Ag8.653 (X63) and that some empty molecules could be stabilized at the cell surface by incubating the cells at 26°C or by adding exogenously eluted peptide material (ULBRECHT et al. 1992). The characterization of antibodies specific for HLA-E (3D12), or for HLA-E and HLA-C (DT9) enabled detection of HLA-E at the cell surface of most cells (BRAUD et al. 1998b; LEE et al. 1998a). Interestingly, in mutant cell line 721.221, HLA-E does not mature within the cell and is not expressed at the cell surface. Because their leader sequences are shortened, peptides capable of binding to HLA-E cannot be generated from the endogenous HLA-E and HLA-F in these cells. Transfection of MHC class I alleles into 721.221 provided peptides capable of binding to HLA-E and restored HLA-E expression at the cell surface (BRAUD et al. 1998b; LEE et al. 1998a). A close correlation was observed between the presence or absence of an HLA leader peptide capable of binding to HLA-E and its cell surface expression, demonstrating that HLA-E cell surface expression depends on binding of a MHC class I leader peptide. Incubation of 721.221 cells at 26°C in the presence of various HLA leader peptides also enhanced HLA-E cell surface expression (LEE et al. 1998a; BORREGO et al. 1998).

Studies of the assembly of HLA-E in the endoplasmic reticulum (ER) revealed that HLA-E associates with the transporter associated with antigen processing (TAP) and the chaperone molecule calreticulin, similarly to classical class I molecules (BRAUD et al. 1998b). Although HLA-E binds signal sequence-derived peptides, HLA-E cell surface expression was found to be TAP and tapasin dependent (BRAUD et al. 1998b; LEE et al. 1998a). The rate of dissociation of HLA-E from

TAP was also enhanced by the introduction in the cytosol of a leader peptide capable of binding to HLA-E (Braud et al. 1998b). It is therefore likely that the leader peptide is released in the cytosol and translocated into the lumen of the ER by the TAP complex before being loaded into HLA-E molecules (Fig. 1).

Previous examples of the presentation of leader sequence peptides in a TAP-dependent manner have also been reported. Interestingly, processing of the pre-prolactin signal sequence generates two fragments, the amino-terminal peptide being released in the cytosol (Lyko et al. 1995). As the leader peptide binding to HLA-E is situated at the N-terminus, it is possible that the MHC class I leader sequences are processed in the same way.

These analyses lead to several conclusions. The cell surface expression of HLA-E is regulated by the binding of a restricted pool of peptides from the leader sequences of MHC class I molecules. HLA-E expression also correlates with the expression of classical class I molecules and HLA-G, because the assembly and loading of peptide into HLA-E requires the same ER resident molecules involved in MHC class I molecules' assembly. Finally, although these HLA-E binding peptides are derived from signal sequences, they are likely released into the cytosol and subsequently translocated by the TAP complex and loaded into HLA-E molecules.

5 HLA-E Is the Ligand for CD94/NKG2 Receptors

The close correlation between the surface expression of HLA-E and that of most other MHC class I molecules suggested a possible role for HLA-E in NK cell-mediated recognition. To test this hypothesis, tetrameric complexes were engineered in which recombinant HLA-E and β2 microglobulin were refolded in vitro with a MHC leader sequence peptide (Braud et al. 1998a). A biotinylation site was added at the C terminus of HLA-E heavy chains truncated of their transmembrane and cytoplasmic domains, so that the refolded complexes could be enzymatically biotinylated. Tetrameric HLA-E complexes were formed by conjugating the refolded, biotinylated HLA-E monomers with phycoerythrine-labeled Extravidin. Tetrameric complexes have slower dissociation rates, making them more suitable for use as staining reagents.

HLA-E tetramers bound to NK cells and a subset of T cells from peripheral blood, suggesting that HLA-E's ligand was an NK cell receptor. Staining was abolished when the PBMC and the tetramer were incubated in the presence of anti-CD94 antibodies, but not when incubated in the presence of antibody specific for NK cell receptors belonging to the immunoglobulin superfamily. Tetramer staining of stable transfectants demonstrated that HLA-E binds to both the CD94/NKG2 A or B inhibitory NK receptor and the CD94/NKG2 C activatory NK receptor, but not to CD94 alone or NKG2 A, B, or C alone. Interestingly, although both CD94 and NKG2 are C-type lectin glycoproteins, their binding to HLA-E was carbohydrate independent, as the soluble HLA-E was produced in *Escherichia coli*.

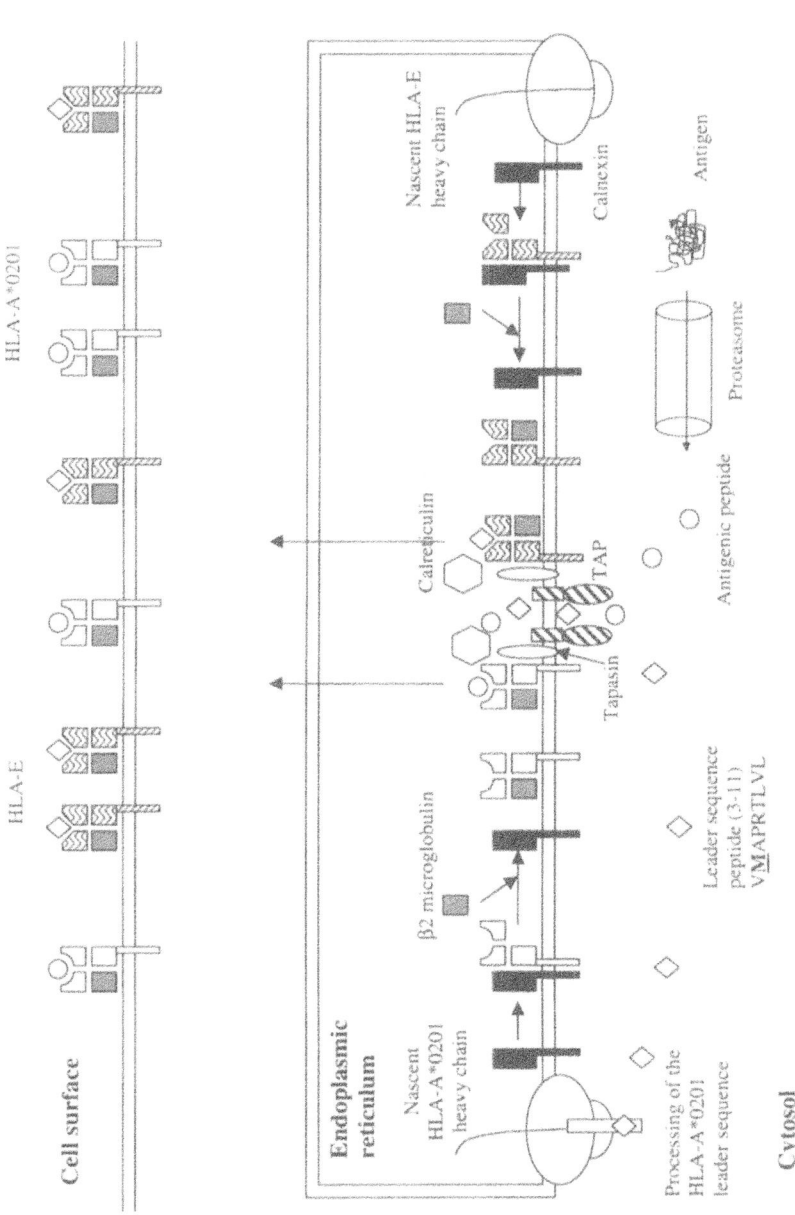

In confirmation, HLA-E tetramer did not bind to cells stably transfected with several immunoglobulin-type NK receptors, and lack of interaction with these receptors was also confirmed in functional assays (LEE et al. 1998b).

The specificity of the inhibitory CD94/NKG2 A receptor was previously unclear. Most, but not all MHC class I molecules could protect a target cell from lysis by CD94/NKG2A$^+$ NK clones (SODERSTROM et al. 1997; PEREZ VILLAR et al. 1997; SIVORI et al. 1996; PHILLIPS et al. 1996; PENDE et al. 1997). That specificity can now be totally correlated with the cell surface expression of HLA-E, suggesting that HLA-E is in fact the ligand for this receptor. Various functional assays proved that HLA-E was mediating inhibition of NK cells through interaction with CD94/NKG2 A receptor (Fig. 2). Chimeric molecules in which a class I leader sequence with a peptide capable of binding to HLA-E were transfected into 721.221 and were capable of protecting the target from lysis (BRAUD et al. 1998a). Addition of class I leader sequence peptides to 721.221 cells or RMA-S cells transfected with HLA-E at 26°C–inorder to stabilize HLA-E at the cell surface–was also sufficient to induce protection (BORREGO et al. 1998; LEE et al. 1998b). It is of note that, in these experiments, both HLA-E alleles (arginine or glycine at position 107) could inhibit NK cell cytotoxicity (BORREGO et al. 1998). Moreover, addition of an HLA-E specific antibody could restore killing of 721.221 cells transfected with HLA alleles possessing a peptide capable of binding to HLA-E in their leader sequences (LEE et al. 1998b). This last experiment also shows that the other HLA alleles are not ligands for CD94/NKG2 A, as the killing was fully restored by the addition of the specific anti-HLA-E antibody.

Engagement of the CD94/NKG2 A receptor by HLA-E induces tyrosine phosphorylation of the NKG2 A subunit and SHP-1 recruitment, suggesting that this receptor uses a pathway common to other ITIM bearing molecules to inhibit cell activation (LE DREAN et al. 1998; CARRETERO et al. 1998).

6 Perspectives

Clearly, HLA-E interaction with CD94/NKG2 receptors is an important and ingenious checkpoint of the immune system in which a non-polymorphic molecule binding a conserved peptide from the leader sequence of highly polymorphic molecules can insure protection from NK-cell-mediated lysis. This is substantiated by the identification of rat and mouse homologues of CD94/NKG2 receptors and HLA-E. A remarkable conservation of the HLA-E peptide binding groove during

---

Fig. 1. HLA-E and MHC class I assembly in human cells. Nascent HLA-E or HLA-A2 heavy chains associate with the chaperone molecule calnexin, which is later displaced by the association of β2 microglobulin. The loading of antigenic peptides into HLA-A2 and MHC leader peptides into HLA-E occurs in the vicinity of the TAP transporter. It involves translocation of the peptides from the cytosol into the lumen of the endoplasmic reticulum through the TAP transporter, and the recruitment of tapasin and calreticulin molecules. HLA-E and HLA-A2 traffic to the cell surface after being loaded with peptides

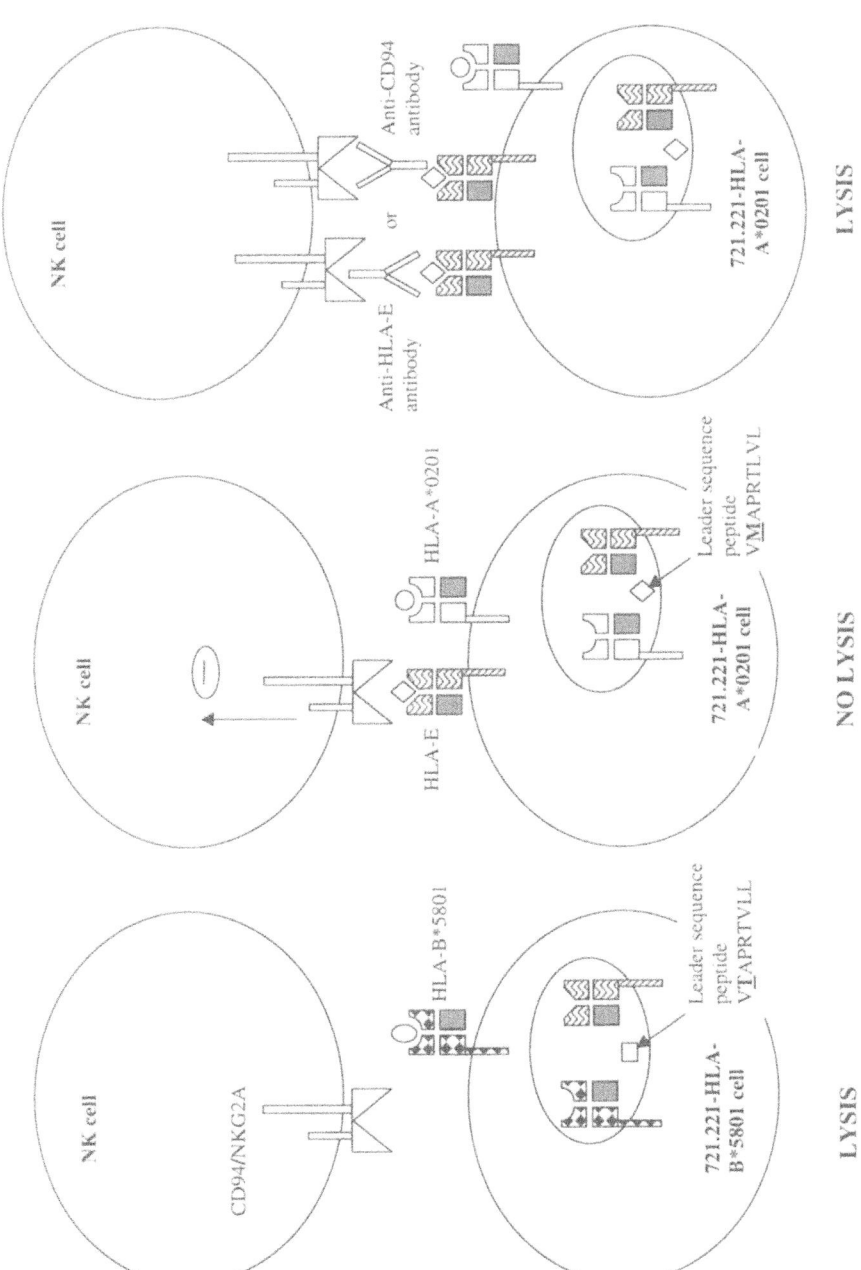

primate evolution also provides further evidence that HLA-E has a critical immunological function (KNAPP et al. 1998).

The potential implications of HLA-E interaction with CD94/NKG2 receptors are various. By using HLA-E, NK cells can monitor the global MHC class I expression on target cells and detect any disruption in the MHC class I assembly and transport pathway. MHC leader peptide availability seems to be a limiting factor for HLA-E expression. Therefore, the loss of some or all HLA alleles can result in reduction or disappearance of HLA-E molecules at the cell surface of target cells, eliminating protection from NK-cell-mediated cytotoxicity. NK cell expression of various receptors recognizing different groups of HLA molecules also ensures, at least partly, the elimination of cells which have lost one HLA allele.

Pathogens have evolved to escape recognition by cytotoxic T cells (CTL) by downregulating their MHC class I molecules. It is likely, however, that they have also found ways to escape HLA-E-CD94/NKG2 A control, either by allowing HLA-E expression while downregulating the other class I molecules, or by producing molecules which could interact directly with CD94/NKG2 A. Further work needs to be done to identify such strategies.

HLA-E is also involved in the regulation of T cell function. Indeed, a significant proportion of αβ and γδ T cells express CD94/NKG2 A receptors. CD94 engagement was shown to facilitate the recruitment of SHP-1 phosphatase to TCR-CD3 complex and to affect phosphorylation of lck and ZAP70 kinases (CARENA et al. 1997). It could also inhibit both antigen-driven TNF release and cytotoxicity on melanoma-specific human T cell clones (LE DREAN et al. 1998). This interference was only observed at suboptimal concentrations of antigen, suggesting that CD94/NKG2 A might regulate the cell activation threshold. Analysis of the phenotype of these T cells expressing NK receptors revealed that they have a memory phenotype and are mainly oligoclonal or monoclonal cell populations, which indicates that they have been activated in vivo. Expression of CD94/NKG2 A on αβ and γδ T cells, similar to the KIRs, could be a way to control T cell activation, which can be of importance with autoreactive T cells. In the case of viral infection or cancer, however, the inhibition of virus or tumor specific CTL by CD94/NKG2 A could lead to the inability of these cells to control diseases. A better characterization of the CD94/NKG2 A expressing T cells is necessary to understand the exact role of this NK receptor on T cells.

Finally, HLA-E is likely to play a crucial role in pregnancy. Indeed, evidence is accumulating that HLA-E is expressed on trophoblast cells along with HLA-G and HLA-C. The majority of NK cells isolated from the maternal uterine mucosa (decidua) are brightly stained with HLA-E tetramer (D Allan, A King, YW Loke, personal communication). Further staining of these NK cells with an anti-NKG2 A

◄──

Fig. 2. Cell surface expression of HLA-E protects target cells from lysis by CD94/NKG2A[+] NK clones. HLA-E cell surface expression depends on the presence of a peptide from the leader sequence of other MHC class I molecules. A threonine at position 2 in the leader peptide disrupts peptide binding to HLA-E whereas a methionine allows binding. Cell surface HLA-E can protect target cells from being lysed by CD94/NKG2A[+] NK cells, while anti-HLA-E and anti-CD94 antibodies reverse this protection

specific antibody confirmed that HLA-E tetramer was binding to the inhibitory CD94/NKG2 A receptor. It is therefore likely that interaction of trophoblast HLA-E with CD94/NKG2 A receptors expressed by decidual NK cells plays an important role in the regulation of placental implantation. Further experiments are in progress to assess its contribution.

Acknowledgments. The authors would like to thank David Allan for critical review of this manuscript. This work was supported by the Medical Research Council.

References

Aldrich CJ, DeCloux A, Woods AS, Cotter RJ, Soloski MJ, Forman J (1994) Identification of a TAP-dependent leader peptide recognized by alloreactive T cells specific for a class Ib antigen. Cell 79: 649–658

Borrego F, Ulbrecht M, Weiss EH, Coligan JE, Brooks AG (1998) Recognition of human histocompatibility leukocyte antigen (HLA)-E complexed with HLA class I signal sequence-derived peptides by CD94/NKG2 confers protection from NK-cell-mediated lysis. J Exp Med 187:813–818

Braud V, Jones EY, McMichael A (1997) The human major histocompatibility complex class Ib molecule HLA-E binds signal sequence-derived peptides with primary anchor residues at positions 2 and 9. Eur J Immunol 27:1164–1169

Braud VM, Allan DSJ, O'Callaghan CA, Soderstrom K, D'Andrea A, Ogg GS, Lazetic S, Young NT, Bell JI, Phillips JH, Lanier LL, McMichael AJ (1998a) HLA-E binds to natural killer cell receptors CD94/NKG2 A, B, and C. Nature 391:795–799

Braud VM, Allan DSJ, Wilson D, McMichael AJ (1998b) TAP- and tapasin-dependent HLA-E surface expression correlates with the binding of an MHC class I leader peptide. Curr Biol 8:1–10

Brooks AG, Posch PE, Scorzelli CJ, Borrego F, Coligan JE (1997) NKG2 A complexed with CD94 defines a novel inhibitory natural killer cell receptor. J Exp Med 185:795–800

Burshtyn DN, Long EO (1997) Regulation through inhibitory receptors: lessons from natural killer cells. Trends. Cell Biol 7:473–479

Cantoni C, Biassoni R, Pende D, Sivori S, Accame L, Pareti L, Semenzato G, Moretta L, Moretta A, Bottino C (1998) The activating form of CD94 receptor complex: CD94 covalently associates with the Kp39 protein that represents the product of the NKG2-C gene. Eur J Immunol 28:327–338

Carena I, Shamshiev A, Donda A, Colonna M, De libero G (1997) Major histocompatibility complex class I molecules modulate activation threshold and early signaling of T cell antigen receptor-γ/δ stimulated by nonpeptidic ligands. J Exp Med 186:1769–1774

Carretero M, Cantoni C, Bellon T, Bottino C, Biassoni R, Rodriguez A, Perez-Villar JJ, Moretta A, Lopez-Botet M (1997) The CD94 and NKG2-A C-type lectins covalently assemble to form a natural killer cell inhibitory receptor for HLA class I molecules. Eur J Immunol 27:563–567

Carretero M, Palmieri G, Liano M, Tullio V, Santoni A, Geraghty DE, Lopez-Botet M (1998) Specific engagement of the CD94/NKG2 A killer inhibitory receptor by the HLA-E class Ib molecule induces SHP-1 phosphatase recruitment to tyrosine-phosphorylated NKG2 A: evidence for receptor function in heterologous transfectants. Eur J Immunol 28:1280–1291

Houchins JP, Lanier LL, Niemi EC, Phillips JH, Ryan JC (1997) Natural killer cell cytolytic activity is inhibited by NKG2-A and activated by NKG2-C. J Immunol 158:3603–3609

Kärre K (1997) How to recognize a foreign submarine. Immunol Rev 155:5–9

Knapp LA, Cadavid LF, Watkins DI (1998) The MHC-E locus is the most well conserved of all known primate class I histocompatibility genes. J Immunol 160:189–196

Lanier LL (1998) NK cell receptors. Annu Rev Immunol 16:359–393

Lanier LL, Corliss B, Wu J, Phillips JH (1998a) Association of DAP12 with activating CD94/NKG2 C NK cell receptors. Immunity 8:693–701

Lanier LL, Corliss BC, Wu J, Leong C, Phillips JH (1998b) Immunoreceptor DAP12 bearing a tyrosine-based activation motif is involved in activating NK cells. Nature 391:703–707

Lazetic S, Chang C, Houchins JP, Lanier LL, Phillips JH (1996) Human natural killer cell receptors involved in MHC class I recognition are disulfide-linked heterodimers of CD94 and NKG2 subunits. J Immunol 157:4741–4745

Le Drean E, Vely F, Olcese L, Cambiaggi A, Guia S, Krystal G, Gervois N, Moretta A, Jotereau F, Vivier E (1998) Inhibition of antigen-induced T cell response and antibody-induced NK cell cytotoxicity by NKG2 A: association of NKG2 A with SHP-1 and SHP-2 protein-tyrosine phosphatases. Eur J Immunol 28:264–276

Lee N, Goodlett DR, Ishitani A, Marquardt H, Geraghty DE (1998a) HLA-E surface expression depends on binding of TAP-dependent peptides derived from certain HLA class I signal sequences. J Immunol 160:4951–4960

Lee N, Llano M, Carretero M, Ishitani A, Navarro F, Lopez-Botet M, Geraghty DE (1998b) HLA-E is a major ligand for the natural killer inhibitory receptor CD94/NKG2 A. Proc Natl Acad Sci USA 95:5199–5204

Lyko F, Martoglio B, Jungnickel B, Rapoport TA, Dobberstein B (1995) Signal sequence processing in rough microsomes. J Biol Chem 270:19873–19878

O'Callaghan CA, Tormo J, Willcox BE, Braud VM, Jakobsen BK, Stuart DI, McMichael AJ, Bell JI, Jones EY (1998) Structural features impose tight peptide binding specificity in the nonclassical MHC molecule HLA-E. Molecular Cell 1:531–541

Olcese L, Cambiaggi A, Semenzato G, Bottino C, Moretta A, Vivier E (1997) Human killer cell activatory receptors for MHC class I molecules are included in a multimeric complex expressed by natural killer cells. J Immunol 158:5083–5086

Pende D, Sivori S, Accame L, Pareti L, Falco M, Geraghty D, Le Bouteiller P, Moretta L, Moretta A (1997) HLA-G recognition by human natural killer cells. Involvement of CD94 both as inhibitory and as activating receptor complex. Eur J Immunol 27:1875–1880

Perez Villar JJ, Melero I, Navarro F, Carretero M, Bellon T, Llano M, Colonna M, Geraghty DE, Lopez Botet M (1997) The CD94/NKG2-A inhibitory receptor complex is involved in natural killer cell-mediated recognition of cells expressing HLA-G1. J Immunol 158:5736–5743

Phillips JH, Chang C, Mattson J, Gumperz JE, Parham P, Lanier LL (1996) CD94 and a novel associated protein (94AP) form a NK cell receptor involved in the recognition of HLA-A, HLA-B and HLA-C allotypes. Immunity 5:163–172

Sivori S, Vitale M, Bottino C, Marcenaro E, Sanseverino L, Parolini S, Moretta L, Moretta A (1996) CD94 functions as a natural killer cell inhibitory receptor for different HLA class I alleles: identification of the inhibitory form of CD94 by the use of novel monoclonal antibodies. Eur J Immunol 26:2487–2492

Soderstrom K, Corliss B, Lanier LL, Phillips JH (1997) CD94/NKG2 is the predominant inhibitory receptor involved in recognition of HLA-G by decidual and peripheral blood NK cells. J Immunol 159:1072–1075

Ulbrecht M, Kellermann J, Johnson JP, Weiss EH (1992) Impaired intracellular transport and cell surface expression of nonpolymorphic HLA-E: evidence for inefficient peptide binding. J Exp Med 176:1083–1090

Inhibitory Role of Murine Ly49 Lectin-like Receptors on Natural Killer Cells

M. Salcedo

1 Introduction

Natural killer (NK) cells are an essential component of the innate immune system, both as cytotoxic effectors and as a source of cytokines that modulate the activity of other cells, including T cells and macrophages. While the molecular mechanisms involved in NK cell activation are not completely understood, significant progress has been made in the characterization of the inhibition of NK cell-mediated lysis. It has become clear that the interaction between target cell MHC class I molecules and specific MHC class I-binding inhibitory receptors on NK cells interrupts the process of NK cell activation (YOKOYAMA 1998a; LANIER 1998). The physiological significance of this type of recognition finds its best explanation in the "missing self" hypothesis (LJUNGGREN and KÄRRE 1990), which suggests that NK cells recognize and kill cells that have lost normal expression of MHC class I molecules, e.g. during viral infections or malignant processes.

    The MHC class I binding inhibitory receptors fall into two distinct structural groups: the Ig-like receptors, which are monomeric type I integral membrane proteins; and the C-type lectin-like receptors, which are disulfide-linked dimeric type II integral membrane proteins. The human KIR family of receptors belong to the first structural group, while the mouse Ly49 family and murine and human

Unité de Biologie Moléculaire du Gène, INSERM U277, Institut Pasteur, 75015 Paris, France
Current address: IDM (Immune Designed Molecules), 75011 Paris, France

CD94 and NKG2 families belong to the lectin-like group of receptors (YOKOYAMA 1998a; LANIER 1998). Ly49 receptors are generally expressed as homodimers whereas CD94/NKG2 proteins associate as heterodimers. cDNAs encoding at least fourteen members of the mouse Ly49 family have been identified (designated Ly49A-N). They have been shown to be located in the mouse NK gene complex (NKC) on distal mouse chromosome 6 (TAKEI et al. 1997; YOKOYAMA 1998a; McQUEEN et al. 1998). Several Ly49 rat homologous genes have also been described (RYAN and SEAMAN 1997). Recently, a disulfide-linked homodimeric molecule with an inhibitory function has been identified on the rat NK complex (NAPER et al. 1998). Sequence analysis has revealed the presence of conserved inhibitory motifs in the cytoplasmic tail of several members of this family of receptors (TAKEI 1997; YOKOYAMA 1998a; McQUEEN et al. 1998). Some of these receptors (Ly49A, Ly49C, Ly49D, Ly49I, and Ly49G2) have been defined by monoclonal antibodies (mAbs), which has facilitated the characterization of their specificities (reviewed in RAULET et al. 1997; TAKEI et al. 1997). For some of these receptors, the specificity has been mapped to the $\alpha1/\alpha2$ domains of the MHC class I molecule (SUNDBÄCK et al. 1998; MATSUMOTO et al. 1998) and to the stalk and carbohydrate recognition domain (CRD) of the Ly49 receptor (TAKEI et al. 1997). Even though polysaccharides impair the interaction between MHC class I ligands and Ly49 receptors, unglycosylated soluble H-2 molecules are able to interact specifically with their correspondent receptors (MATSUMOTO et al. 1998). Experimental data have suggested that the MHC class I-loaded peptide is not specifically recognized by murine NK inhibitory receptors, although it is required for appropriate folding and expression of a conformational determinant (CORREA and RAULET 1995; ORIHUELA et al. 1996). Recent studies, however, have shown that certain peptides that bind equally well to H-2K^b molecules differ in their protective capacity against NK1.1$^+$/Ly49C$^+$ cells (Lars Franksson, personal communication)

2 Inhibitory Signals Mediated by Ly49 Lectin-like Receptors and Related Recruitment of Tyrosine Phosphatases

It has become evident that a broad spectrum of cell functions are regulated by both activating and inhibitory stimuli. Immune cells can generate such signals through cleverly designed receptors that, by themselves or in association with other proteins, are able to direct the intracellular signaling pathways toward either positive or negative responses. Conserved intracytoplasmic sequences on these receptors or on associated proteins have been identified. These conserved motifs fall into two homologous groups: The immunoreceptor tyrosine-based activation motifs, or ITAMs (consensus amino acid sequence YxxL-x$_{(6-8)}$-YxxL, single letter amino acid code and x representing any amino acid), and, the immunoreceptor tyrosine-based inhibitory motifs, or ITIMs (consensus amino acid sequence I/VxYxxL/V) (LEIBSON 1997). When the ligand-binding component of the receptor is engaged, the cytoplasmic tyrosine residues of the ITIM or the ITAM are phosphorylated by src-family tyrosine kinases. ITIM phosphorylation is followed by recruitment of molecules containing tyrosine phophatases such as SHP-1, SHP-2, or SHIP, while ITAM phosphorylation leads to the recruitment of non-receptor-related src- and syk/zap-70-family tyrosine kinases (LEIBSON 1997; VIVIER and DAEBRON 1997).

Ly49 inhibitory receptors have been shown to possess ITIM-like sequences. Several of these receptors, including Ly49A, Ly49C, and Ly49G2, have been shown to be phosphorylated following pervanadate stimulation or mAb-induced receptor ligation. It has also been shown that SHP-1 coprecipitates with Ly49A and Ly49G2 following receptor phosphorylation (MASON et al. 1997). The importance of this tyrosin phosphatase was supported by studies using SHP-1-deficient moth-eaten mice, in which Ly49 inhibitory function is impaired, although not completely abolished. Furthermore, tyrosine substitution within the proposed SHP-1 binding motif abrogates receptor phosphorylation (RYAN and SEAMAN 1997). Ly49 has also been shown to recruit SHP-2 (OLCESE et al. 1996). The tyrosine phosphatases SHP-1 and SHP-2 have an enzymatic activity that blocks the activation cascade at an early stage. The substrate for this enzyme during NK cell inhibition is not known. Interestingly, engagement of KIRs disrupts the interaction between phospholipase Cγ-1 (PLCγ-1) and the adapter protein pp36, blocking target cell-induced activation of phospholipase C. In vitro, pp36 serves as a substrate for the NK cell receptor-associated SHP-1, suggesting that engagement of an NK cell inhibitory receptor by its MHC class I ligand prevents pp36-PLCγ-1 interactions by inducing dephosphorylation of pp36. Alternatively, proximal src-family or syk-family PTK could be the physiological targets for SHP-1 (WEBER et al. 1998; reviewed in LEIBSON 1997).

Ly49D receptor has been implicated in NK cell activation. This receptor, as well as Ly49H, Ly49L, Ly49K, and Ly49N, lack ITIMs. These truncated receptors contain a charged amino acid in the transmembrane domain, a shared characteristic with human CD94/NKG2 and short KIR receptors that have been shown to be involved in the transmission of activating signals; however, these molecules also lack ITAMs (LANIER 1998). Recently, a signal transduction molecule called DAP12 has been demonstrated to couple to Ly49D and Ly49H (SMITH et al. 1998) as well as to the human activating receptors CD94/NKG2C and KIR2DS2 (LANIER et al. 1998a,b). DAP12 is a disulphide-linked homodimer that contains an ITAM-like sequence and binds ZAP-70 and syk protein tyrosine kinases upon phosphorylation. Since DAP12 has a negatively charged residue in its transmembrane portion, it probably associates noncovalently with the NK cell receptor, helped by the interaction of transmembrane residues with an opposite charge (LANIER et al. 1998a). These data suggest that isoforms of NK cell receptors lacking ITIMs appear to transmit intracellular signals through an associated molecule as an activating multimeric complex.

3 Regulation of the Expression and the Function of Ly49 Inhibitory Receptors

The absence of autoaggression by NK cells in normal individuals implies the existence of a selection process for Ly49-defined NK cell subpopulations that ensures the expression of useful inhibitory receptors and the discarding of harmful ones. Several experimental models have shown that host MHC class I molecules shape the Ly49 receptor pattern of expression (OLSSON et al. 1995; HELD et al. 1996a; SALCEDO et al. 1997). However, NK cells from a given mouse strain can co-express Ly49 receptors with both syngeneic and allogeneic specificities.

The molecular processes involved in the acquisition of an NK cell repertoire are not yet fully understood. In contrast to T cells, NK cells expressing MHC class I-binding receptors mature in mice lacking normal expression of H-2 products. These mice, however, present a moderate augmentation in the numbers of NK1.1$^+$/ Ly49$^+$ cells as well as higher surface expression of these receptors on NK cells (SALCEDO et al. 1997, 1998). Several independent studies have demonstrated that Ly49 receptors are downregulated in mice expressing their specific ligands (OLSSON et al. 1995; GOSSELIN et al. 1997; SALCEDO et al. 1997, 1998). Experimental evidence has suggested that this is a postranscriptional event (FAHLÉN et al. 1997; HELD and RAULET 1997a) and that surface levels of Ly49 are subject to significant changes after in vitro culture with IL-2 (SALCEDO et al. 1998). Interestingly, this calibration to self MHC class I molecules has been shown to alter NK cell specificity. Ly49-low-expressing NK cells appear to acquire a greater capacity to detect quantitative reductions of its ligand on target cells when compared with Ly49-high-expressing cells (OLSSON et al. 1997).

MHC class I-deficient mice should be prone to autoaggression by their own NK cells. However, these mice do not present such pathology. Analyses of fetal liver irradiation chimeras and transgenic mice displaying a mosaic expression of MHC class I products demonstrated that tolerance is dominantly induced by MHC class I deficient cells (WU and RAULET 1997; JOHANSSON et al. 1997; ANDERSSON et al. 1998). In these in vivo models, however, the generalized downregulation of Ly49 receptors specific for the MHC class I mosaic gene product is not compatible with a tolerance mechanism based only on changes in the pattern of expression of Ly49 receptors. It is possible that tolerance is actively regulated by continuous suppression of not yet characterized activating receptors rather than by overexpression of inhibitory ones. In this regard, it is interesting to mention that MHC class I positive NK cells from these mosaic mice lose their tolerance to autologous MHC class I deficient targets after in vitro culture with IL-2 (JOHANSSON et al. 1997). Similarly, NK cells derived from MHC class I-deficient mice tend to lose self-tolerance upon in vitro culture with IL-2 (SALCEDO et al. 1998a).

The steps that determine selection, education and maturation of NK cell subsets are still unclear. It has been shown that after in vivo transfer, NK cells not expressing specific Ly49 receptors can give rise to NK cells that do, and cells expressing one given Ly49 receptor can give rise to cells expressing others (DORFMAN and RAULET 1998). NK cells containing two Ly49 alleles of the same gene, tend to express only one of them. In certain occasions both alleles are expressed. However, this co-expression appears to be unrelated (HELD and RAULET 1997b; HELD and KUNZ 1998). Thus, these studies suggest that activation of the expression of Ly49 receptor encoding genes occurs successively and follows stochastic mechanisms. However, certain rules that for the moment remain somewhat obscure must build a self-tolerant and stringent NK cell repertoire with a restricted number of cells expressing multiple self-specific Ly49 receptors. A more complete panel of mAbs and functional studies assessing the killing capacity of different subpopulation of NK cells, as well as the generation of a model that allows generating Ly49$^+$ NK cells from immature precursor cells in vitro, are required to further characterize the selection of an appropriate NK cell repertoire by host MHC class I and to define the rules that govern these processes.

4 Other Murine Lectin-like Receptors Involved in NK Cell-mediated Lysis Inhibition

In an attempt to identify Ly49 homologous genes on human NK cells, several genes encoding lectin-like receptors were identified. These proteins constitute a second type of human MHC class I inhibitory receptors and are expressed as heterodimers named CD94/NKG2. The NKG2 family of proteins encodes at least four different type II transmembrane molecules (NKG2A, NKG2C, NKG2E, and NKG2D/F). Since CD94 molecules present a short cytoplasmic domain, the intracellular signal delivered through this heterodimeric receptor can apparently be derived from the cytoplasmic domain of NKG2. NKG2A and NKG2B (an alternative splicing version of NKG2A) contain ITIMs and are inhibitory whereas NKG2C does not contain an ITIM and is stimulatory (reviewed in LANIER 1998). cDNAs encoding rat and mouse CD94 and NKG2 genes have been lately identified (reviewed in YOKOYAMA 1998b).

Recent studies have demonstrated that human CD94/NKG2 inhibitory receptors bind the nonclassical MHC class I molecule HLA-E. For cell surface expression, HLA-E requires binding of the leader peptide of a classical MHC class I molecule (BRAUD et al. 1998). In retrospect, this observation may explain why the specificity for the CD94/NKG2A receptor was considered promiscuous with implicated binding specificities to HLA-A, -B, -C, and -G molecules. The class Ib molecule Qa-1, which is widely expressed in the mouse, shares several features with HLA-E. The most striking one is the changes at position 143 and 147, where conserved threonine and tryptophan are replaced by serine, potentially enlarging the F pocket. This changes may explain the restricted peptide repertoire displayed by both Qa-1 and HLA-E, which is largely comprised of leader sequences from other MHC class I molecules (SOLOSKI et al. 1995; BRAUD et al. 1998). One of the common functions of these conserved molecules may be to serve as a ligand for (an) NK cell receptor(s). We have seen that Qa-1 in the form of soluble tetramers binds to a distinct subset of fresh or IL-2-activated NK1.1$^+$/CD3$^-$ splenocytes, and that binding does not correlate with expression of specific Ly49 receptors with known specificity for classical MHC class I molecules. Binding occurs whether NK cells have evolved in an MHC class I expressing environment or in an MHC class I deficient one. Recent studies have shown that the Qa-1 tetramers bind to mouse CD94/NKG2A heterodimer. This interaction transmits an inhibitory signal (SALCEDO et al. 1998b; VANCE et al. 1998). The NKG2D gene identified in mouse does not contain ITIMs and contains charged residues in its transmembrane domain that are incompatible in the heterodimeric configuration with the charged residue present in CD94 (Ho et al. 1998).

5 Concluding Remarks

In spite of the overwhelming amount of knowledge acquired in recent years about the modulation of NK cell activity, this field is constantly expanding. Although the regulation of the specificity of NK cells has started to be unravelled, there are questions remaining to be answered. Which signal threshold determines whether an NK cell should kill? How can the co-existence of both activating and inhibitory MHC class I-binding NK receptors be reconciled? What are the in vivo implications of what is known about NK cell-mediated target recognition?

One can think of several possibilities to explain how an NK cell reaches a killing decision. It is important to know that the current available data suggest that the MHC-recognizing receptors exert their inhibitory effect on NK cellular activation only when they are coaggregated with activating receptors. This would facilitate the physical contact between NK cell receptor-associated tyrosine phosphatases and their substrates in the activation complex (LEIBSON 1997; VIVIER and DAEBRON 1997). It may be that activating receptors cooperate with inhibitory ones by recruiting tyrosine kinases to allow phosphorylation of the cytoplasmic ITIMs. On the other hand, a single NK cell could express an activating receptor specific for one MHC class I allele and an inhibitory receptor specific for another one. This would allow an NK cell to be activated when exposed to target cells that have lost a single MHC class I allele. However, NK cell-mediated recognition cannot be explained by these simplistic explanations. Attempts to identify a single triggering receptor for natural cytotoxicity have been unsuccessful. In fact, several different molecules, including lectin-like proteins, adhesion, and costimulatory molecules, are able to trigger NK cell activation upon specific engagement (reviewed in LEIBSON 1997; YOKOYAMA 1998a; LANIER 1998). The rapid expansion of this field will surely lead to the identification and molecular characterization of novel molecules involved in the signaling events that take place during NK cell activation (and inhibition), which would help complete this exciting puzzle.

Several studies have integrated the regulation of the NK cell receptor-mediated target recognition and its importance in specific in vivo situations. For example, the fate of bone marrow allografts have been shown to be significantly influenced by certain Ly49 subpopulations. It has been demonstrated that depletion of certain subsets of NK cells may lead to acceptance or rejection of bone marrow grafts, depending on whether the depleted subpopulation expresses an inhibitory or and activating NK cell receptor (RAZIUDDIN et al. 1998). Presence of inhibitory receptors on decidual NK cells recognizing MHC class I alleles expressed by trophoblasts may provide a useful system to protect the fetus from recognition by the mother 27s NK cells or to stimulate NK cells to produce relevant cytokines related to pregnancy (KING et al. 1997). At last, inhibitory as well as activating receptors have been shown to be present on other immune cells. In the human, T cell activity has been demonstrated to be affected by the presence of KIRs (DANDREA and LANIER 1998). Ly49A has been reported to be expressed in $CD3^+$ $\alpha\beta$ intestinal intra-epithelial T lymphocytes (ROLAND and CAZENAVE 1992). The role of Ly49 NK cell receptors on murine T cells has not been widely explored. However, studies in Ly49A transgenic mice have shown that this receptor prevents T-cell in vitro proliferative response to stimulator cells of allogeneic origin (HELD et al. 1996b).

Generation of reagents such as soluble receptors and ligands is crucial to assess molecular aspects of the modulation of NK cell activation/inhibition equilibrium. The establishment of experimental models where regulation of the expression of inhibitory receptors can be modified and followed will also constitute an important source of information in the study of the events that determine NK cell specificity.

Acknowledgments. I thank Dr. H.G. Ljunggren for critical reading of this article. The author is currently on a fellowship from the INSERM, France.

References

Andersson M, Freland S, Johansson M, Wallin R, Sandberg JK, Chambers BJ, Christensson B, Lendhal U, Lemieux S, Salcedo M, Ljunggren HG (1998) MHC class I mosaic mice reveal insights into control of Ly49C inhibitory receptor expression in NK cells. J Immunol In press

Braud VM, Allan DSJ, Wilson D, McMichael AJ (1998) Tap- and tapasin-dependent HLA-E surface expression correlates with the binding of an MHC class I leader peptide. Curr Biol 8:1–10

Correa I, Raulet DH (1995) Binding of diverse peptides to MHC class I molecules inhibits target cell lysis by activated natural killer cells. Immunity 2:61–71

DAndrea A, Lanier LL (1998) Killer cell inhibitory receptor expression by T cells. Curr Top Microbiol Immunol 230:25–39

Dorfman JR, Raulet DH (1998) Acquisition of Ly49 receptor expression by developing natural killer cells. J Exp Med 187:609–618

Fahlén L, Khoo NKS, Daws MR, Sentman CL (1997) Location-specific regulation of transgenic Ly49A by major histocompatibility complex class I molecules. Eur J Immunol 27:2057–2065

Gosselin P, Lusignan Y, Brennan J, Takei F, Lemieux S (1997) The NK2.1 receptor is encoded by Ly-49C and its expression is regulated by MHC class I alleles. Int Immunol 9:533–540

Held W, Kunz B (1998) An allele-specific, stochastic gene expression process controls the expression of multiple Ly49 family genes and generates a diverse, MHC-specific NK cell receptor repertoire. Eur J Immunol 28:2407–2416

Held W, Raulet DH (1997a) Ly49A transgenic mice provide evidence for a major histocompatibility complex-dependent education process in natural killer cell development. J Exp Med 185:2079–2088

Held W, Raulet DH (1997b) Expression of the Ly49A gene in murine natural killer cell clones is predominantly but not exclusively mono-alleleic. Eur J Immunol 27:2876–2884

Held W, Dorfman JR, Wu MF, Raulet DH (1996a) Major histocompatibility complex class I-dependent skewing of the natural killer cell Ly49 receptor repertoire. Eur J Immunol 26:2286–2292

Held W, Cado D, Raulet DH (1996b) Transgenic expression of the Ly49A natural killer cell receptor confers class I major histocompatibility complex (MHC)-specific inhibition and prevents bone marrow allograft rejection. J Exp Med 184:2037–2041

Ho EL, Heusel JW, Brown MG, Matsumoto K, Scalzo AA, Yokoyama WM (1998) Murine NKG2D and CD94 are clustered within the natural killer complex and are expressed independently in natural killer cells. Proc Natl Acad Sci USA 95:6320–6325

Johansson MH, Bieberich C, Jay G, Kärre K, Höglund P (1997) Natural killer cell tolerance in mice with mosaic expression of major histocompatibility complex class I transgene. J Exp Med 186:353–364

King A, Loke YW, Chaouat G (1997) NK cells and reproduction. Immunol Today 18:64–66

Lanier LL (1998) NK cell receptors. Annu Rev Immunol 16:359–393

Lanier LL, Corliss BC, Wu J, Leong C, Phillips JH (1998a) Immunoreceptor DAP12 bearing a tyrosine-based activation motif is involved in activating NK cells. Nature 391:703–707

Lanier LL, Corliss BC, Wu J, Phillips JH (1998b) Association of DAP12 with activating CD94/NKG2C NK cell receptors. Immunity 8:693–701

Leibson PJ (1997) Signal transduction during natural killer cell activation: inside the mind of a killer. Immunity 6:655–661

Ljunggren HG, Kärre K (1990) In search of the missing self: MHC molecules and NK cell recognition. Immunol Today 11:237–244

Mason LH, Gosselin P, Anderson SK, Fogler WE, Ortaldo JR, McVicar DW (1997) Differential tyrosine phosphorylation of inhibitory versus activating Ly-49 receptor proteins and their recruitment of SHP-1 phosphatase. J Immunol 159:4187–4196

Matsumoto N, Ribaudo RK, Abastado JP, Margulies DH, Yokoyama WM (1998) The lectin-like NK cell receptor Ly-49A recognizes a carbohydrate-independent epitope in its MHC class I ligand. Immunity 8:245–254

McQueen KL, Freeman JD, Takei F, Mager DL (1998) Localization of five new Ly49 genes, including three closely related to Ly49C. Immunogenetics 48:174–183

Naper C, Ryan JC, Nakamura MC, Lambracht D, Rolstad B, Vaage JT (1998) Identification of an inhibitory MHC receptor on alloreactive rat natural killer cells. J Immunol 160:219–224

Olcese L, Lang P, Vely F, Cambiaggi A, Marguet D, Blery M, Hippen KL, Biassoni R, Moretta A, Moretta L, Cambier JC, Vivier E (1996) Human and mouse killer-cell inhibitory receptors recruit PTP1C and PTP1D protein tyrosine phosphatases. J Immunol 156:4531–4534

Olsson MY, Kärre K, Sentman CL (1995) Altered phenotype and function of natural killer cells expressing the major histocompatibility complex receptor Ly-49 in mice transgenic for its ligand. Proc Natl Acad Sci USA 92:1649–1653

Olsson-Alheim MY, Salcedo M, Ljunggren HG, Kärre K, Sentman CL (1997) NK cell receptor calibration: effects of MHC class I induction on killing by Ly49A^{high} and Ly49A^{low} NK cells. J Immunol 159:3189–3194

Orihuela M, Margulies DH, Yokoyama WM (1996) The natural killer cell receptor Ly-49A recognizes a peptide-induced conformational determinant on its major histocompatibility complex class I ligand. Proc Natl Acad Sci USA 93:11792–11797

Raulet DH, Held W, Correa I, Dorfman JR, Wu MF, Corral L (1997) Specificity, tolerance and developmental regulation of natural killer cells defined by expression of class I-specific Ly49 receptors. Immunol Rev 155:41–52

Raziuddin A, Longo DL, Mason L, Ortaldo JR, Bennett M, Murphy WJ (1998) Differential effects of the rejection of bone marrow allografts by the depletion of activating versus inhibiting Ly-49 natural killer cell subsets. J Immunol 160:87–94

Roland J, Cazenave PA (1992) Ly-49 antigen defines an αβ TCR population in i-IEL with an extrathymic maturation. Int Immunol 4:699–706

Ryan JC, Seaman WE (1997) Divergent functions of lectin-like receptors on NK cells. Immunol Rev 155:79–89

Salcedo M, Andersson M, Lemieux S, Van Kaer L, Chambers BJ, Ljunggren HG (1998a) Fine tuning of natural killer cell specificity and maintenance of self tolerance in MHC class I deficient mice. Eur J Immunol 28:1315–1321

Salcedo M, Bousso P, Ljunggren HG, Kourilsky P, Abastado JP (1998b) The Qa-1[b] molecule binds to a large subpopulation of murine NK cells. Eur J Immunol 28:4356–4361

Salcedo M, Diehl AD, Olsson-Alheim MY, Sundbäck J, Van Kaer L, Kärre K, Ljunggren HG (1997) Altered expression of Ly49 inhibitory receptors on natural killer cells from MHC class I-deficient mice. J Immunol 158:3174–3180

Smith KM, Wu J, Bakker ABH, Phillips JH, Lanier LL (1998) Ly-49D and Ly-49H associate with mouse DAP12 and form activating receptors. J Immunol 161:7–10

Soloski MJ, DeCloux A, Aldrich CJ, Forman J (1995) Structural and functional characteristics of the class Ib molecule, Qa-1. Immunol Rev 147:67–89

Sundbäck J, Nakamura MC, Waldenstrf6m M, Niemi EC, Seaman WE, Ryan JC, Kärre K (1998) The α-2 domain of H-2D[d] restricts the allelic specificity of the murine NK cell inhibitory receptor Ly-49A. J Immunol 160:5971–5978

Takei F, Brennan J, Mager DL (1997) The Ly-49 family: genes, proteins and recognition of class I MHC. Immunol Rev 155:67–77

Vance RE, Kraft JR, Altman JD, Jensen PE, Raulet DH (1998) Mouse CD94/NKG2A Is a Natural Killer Cell Receptor for the Nonclassical Major Histocompatibility Complex (MHC) Class I Molecule Qa-1b. J Exp Med 188:1841–1848

Vivier E, Daebron M (1997) Immunoreceptor tyrosine-based inhibition motifs. Immunol Today 18: 286–291

Weber JR, Orstavik S, Torgersen KM, Danbolt NC, Berg SF, Ryan JC, Tasken K, Imboden JB, Vaage JT (1998) Molecular cloning of the cDNA encoding pp36, a tyrosine-phosphorylated adaptor protein selectively expressed by T cells and natural killer cells. J Exp Med 187:1157–1161

Wu MF, Raulet DH (1997) Class I-deficient hemopoietic cells and nonhemopoietic cells dominantly induce unresponsiveness of natural killer cells to class I-deficient bone marrow cell grafts. J Immunol 158:1628–1633

Yokoyama WM (1998a) Natural killer cell receptors. Curr Opin Immunol 10:298–305

Yokoyama WM (1998b) HLA class I specificity for natural killer cell receptor CD94/NKG2A: two for one in more ways than one. Proc Natl Acad Sci USA 95:4791–4794

gp49: An Ig-like Receptor with Inhibitory Properties on Mast Cells and Natural Killer Cells

N. Wagtmann

1 Introduction

The activation of human natural killer (NK) cells is under tight negative control of killer cell Ig-like receptors (KIR) specific for HLA class I. Recently, several new receptor families were identified which share structural features with KIR but are expressed in various different cell types of the immune system. One of these receptor families, the gp49 glycoproteins, are expressed on mouse mast cells and NK cells where they may deliver dominant inhibitory signals participating in the regulation of cellular effector functions.

2 Control of NK and Mast Cell Activation

Mast cells and NK cells are potent mediators of early, innate immunity against various pathogens. For example, mast cells are crucial for resistance to some bacterial infections (ECHTENACHER et al. 1996; MALAVIYA et al. 1996). NK cells can directly kill target cells, including tumor cells and cells infected by some viruses. In addition to their direct involvement in elimination of pathogens, mast and NK cells also regulate other cells of the immune system by secreting cytokines such as TNF-α and IFN-γ (BIRON 1997; TRINCHIERI 1995).

Centre d'Immunologie, INSERM/CNRS de Marseille-Luminy, Case 906, 13288 Marseille, cedex 09, France

Whereas it is well established that NK and mast cells are essential for immunity against certain pathogens, the molecular mechanisms controlling their activation are not fully understood. The best-characterized activating receptors on mast and NK cells are the FcεRI and FcγRIII, which can trigger mast cell degranulation and NK-mediated cytotoxicity upon binding to their ligands, IgE and IgG, respectively (DAËRON 1997). However, both cell types can also be activated in the absence of Ig. In NK cells, this may occur following cross-linking of various activating receptors or adhesion molecules. Signaling through these different activating receptors on NK cells is tightly regulated by inhibitory receptors. NK cells express multiple families of inhibitory receptors specific for distinct major histocompatibility complex (MHC) class I molecules (reviewed by LANIER 1998). These receptor families exhibit a rich diversity of structures, comprising type II transmembrane molecules of the C-type lectin family (the mouse Ly49 and mouse and human NKG2/CD94) and type I transmembrane proteins belonging to the Ig superfamily (the human KIR). Recognition of self-MHC class I molecules on target cells by these receptors results in a negative signal that dominates over activation to prevent NK-mediated killing. Conversely, target cells that fail to express normal MHC class I molecules, such as tumor cells and virus-infected cells, may be killed by NK cells (LJUNGGREN and KÄRRE 1990). Thus, inhibitory receptors specific for MHC class I molecules play an important role in the control of NK cell activation.

Despite their very different overall structures the KIR, Ly-49, and NKG2 inhibitory receptors all employ the same pathway for transducing negative signals (reviewed by BURSHTYN and LONG 1997). The cytoplasmic tails of these molecules share a highly conserved tyrosine phosphorylation motif consisting of the amino acid sequence (Q/E)EVTY(A/T)QL. In addition, KIR and NKG2 molecules possess a second tyrosine phosphorylation motif with the consensus sequence V/IxYxxL (where x is any amino acid). Tyrosine phosphorylation of these motifs leads to recruitment and activation of the cytoplasmic tyrosine phosphatase SHP-1, resulting in inhibition of NK-mediated lysis (BURSHTYN and LONG 1997).

3 gp49 Glycoproteins

A small family of glycoproteins called gp49, which have relatively low sequence homology to human KIR, are expressed on the surface of mouse mast cells and NK cells (ARM et al. 1991; WANG et al. 1997; ROJO et al. 1997). The gp49 molecules have also been detected in peritoneal macrophages and mononuclear phagocytes (LEBLANC and BIRON 1984). Two isoforms, gp49A and gp49B, have very homologous extracellular portions, with 89% amino acid identities in their two extracellular Ig-like domains (CASTELLS et al. 1994). In contrast, their cytoplasmic tails are quite different. Whereas gp49B has a cytoplasmic tail of 74 amino acids containing two tyrosine phosphorylation motifs, gp49A has a short, truncated tail of only 42 amino acids which lacks tyrosine-based signaling motifs (CASTELLS et al.

1994). These structural features suggests that gp49A and B have different signaling capabilities but may share specificity for the same, as yet unidentified ligand. However, the possibility that the two isoforms of gp49 bind different ligands should not be excluded. In the related KIR molecules, single amino acids can determine the specificity for distinct HLA class I allotypes (WINTER and LONG 1997). The mRNAs encoding both gp49A and B have been detected in populations of mast and NK cells (WANG et al. 1997; ROJO et al. 1997). By FACS analysis, essentially all NK cells express gp49 glycoproteins on their surface (WANG et al. 1997), but due to the lack of specific serological reagents, it has not been determined whether individual cells express gp49A or gp49B, or co-express both forms. The gp49A and B molecules are encoded by two distinct genes located head to tail, and separated by 4.4kb (KUROIWA et al. 1998).

The two tyrosine phosphorylation motifs in the cytoplasmic tail of gp49B are highly related to those present in the tail of KIR. Most strikingly, the tail of gp49B contains a sequence, QDVTYAQL, almost identical to the QEVTYAQL present in KIR. This motif, as well as the second IVYAQV tyrosine phosphorylation motif in the tail of gp49B, conforms to the established I/VxYxxL/V motif for binding to and activating the cytoplasmic tyrosine phosphatase SHP-1 (BURSHTYN et al. 1997). In KIR, phosphorylation of the tyrosines in these motifs is essential for generation of the negative signal leading to inhibition of NK-mediated killing (ROJO et al. 1997).

The presence of KIR-related tyrosine phosphorylation motifs in the cytoplasmic tail of gp49B suggested that it may function as an inhibitory receptor. This hypothesis is supported by in vitro data obtained by two experimental strategies. In the first, a recombinant receptor was constructed where the cytoplasmic tail of gp49B replaced the cytoplasmic tail of a KIR specific for HLA-Cw3 (ROJO et al. 1997). This KIR/gp49B chimeric receptor was transiently expressed on human or mouse NK cells. The KIR/gp49B receptor delivered a strong inhibitory signal which completely prevented NK-mediated killing of target cells expressing HLA-Cw3, the specific KIR ligand. Even very strong stimulatory signals generated in antibody-dependent cellular cytotoxicity assays were dominantly inhibited by the cytoplasmic tail of gp49B. The inhibitory signals generated by the cytoplasmic tail of gp49B in these experiments were at least as strong as those delivered through a wild-type KIR tested in parallel (ROJO et al. 1997). In the second experimental approach, antibody-mediated co-cross-linking of gp49 to the FcεRI receptor inhibited FcεRI-triggered degranulation in mast cells (KATZ et al. 1996). These in vitro experiments demonstrated that gp49B can function as an inhibitory receptor in NK and mast cells.

The conservation of tyrosine phosphorylation motifs in KIR and gp49B, and the finding that both the KIR/gp49B chimera and a wild-type KIR inhibited cytotoxicity by either human or mouse NK cells, suggests that both molecules may transduce negative signals through the same conserved pathway. Signaling through KIR relies critically on recruitment and activation of SHP-1, as demonstrated in functional experiments using dominant negative mutants of SHP-1 (BURSHTYN et al. 1996; BINSTADT et al. 1996). Recently, tyrosine phosphorylated synthetic peptides containing either of the conserved motifs in gp49B were shown to bind to

SHP-1 and SHP-2 in mast cell extracts, indicating that gp49B may also transduce negative signals through SHP-1 and possibly SHP-2 (KUROIWA et al. 1998). Surprisingly, the peptide containing the QDVTYAQL motif of gp49B also bound to the inositol phosphatase SHIP in mast cell extracts (KUROIWA et al. 1998). In contrast, the QEVTYAQL sequence from KIR selectively bound to SHP-1 but not SHIP in extracts from T, NK or B cells (BURSHTYN et al. 1997). Functional experiments in NK cells and B cells demonstrated that inhibition through KIR involves SHP-1 but not SHIP, whereas negative signaling through FcγRIIb requires SHIP but not SHP-1 (GUPTA et al. 1997; ONO et al. 1997). Although the functional role of SHIP in gp49B-mediated signaling has not been tested, the binding of gp49B-derived phosphopeptides to both SHP-1 and SHIP raises the possibility that gp49B may transduce negative signals through either phosphatase, perhaps expanding the range of activation signals that may be targeted for inhibition by ligation of gp49B.

The structural features and the functional and biochemical experiments described above strongly suggest that gp49B is an inhibitory receptor which may negatively regulate the activation of mast cells and NK cells. As gp49 expression has also been detected in some macrophages and monocytes, it may participate in the regulation of these cell types as well. Thus, mast cells, NK cells, and possibly monocytes/macrophages may all be regulated by (a) common inhibitory ligand(s). It will be of great interest to determine the identity of this ligand and how it contributes to the regulation of cells expressing gp49. Since the gene encoding gp49B apparently is non-polymorphic (WANG et al. 1997), the ligand for gp49 may be non-variable. In contrast, the KIR and Ly-49 families of inhibitory NK receptors comprise diverse receptors specific for polymorphic MHC class I allotypes. gp49 is expressed on all resting or activated mouse-spleen-derived NK cells, including some clones which lack Ly-49 molecules (WANG et al. 1997), suggesting that this receptor may play an important role in preventing cellular activation, perhaps during certain stages of development or in anatomical locations where activation of NK or mast cells could have detrimental consequences.

Mast cells, NK cells, and monocytes/macrophages all can be inhibited through additional receptors besides gp49. NK cells express KIR, Ly-49, and CD94/NKG2 (LANIER 1998) and mast cells express FcγRIIb, MAFA, and CD81 (DAËRON et al. 1995; GUTHMANN et al. 1995; FLEMMING et al. 1997). Monocytes and macrophages express members of the recently identified ILT/LIR/MIR, the p91/PIR, and LAIR families of receptors which contain Ig-like domains related to those in gp49 and KIR and cytoplasmic tails with tyrosine phosphorylation motifs specific for SHP-1 (SAMARIDIS and COLONNA 1997; COSMAN et al. 1997; BORGES et al. 1997; ARM et al. 1997; WAGTMANN et al. 1997; HAYAMI et al. 1997; KUBAGAWA et al. 1997; MEYAARD et al. 1997). The members of these receptor families are expressed in additional, often overlapping, subsets of cells of the immune system, as discussed in other reviews in this issue. The discovery of inhibitory receptors expressed in a broad range of cell types, and the finding that some cell types such as mast and NK cells express multiple types of inhibitory receptors suggests that inhibition is an important, widespread mechanism of regulation in the immune system.

Human homologues of gp49 have not been identified yet. Sequence comparisons of individual Ig domains indicates that the closest relatives of gp49 are the mouse p91/PIR and the human ILT/LIR/MIR families of proteins which have six and four Ig-like domains, respectively. Such sequence comparisons suggest that gp49, p91/PIR, KIR, ILT/LIR/MIR, and LAIR are evolutionarily related molecules which have diverged in the extracellular domains to recognize different ligands while retaining the same tyrosine phosphorylation motifs for initiating negative signaling via SHP-1. Whereas the genes encoding the KIR, ILT/LIR/MIR, and LAIR molecules are on human chromosome 19q13.4, and p91/PIR is located syntenically on mouse chromosome 7 (WAGTMANN et al. 1997), the genes encoding gp49 are on mouse chromosome 10 at cytogenetic band B4 (KUROIWA et al. 1998).

The function of gp49A is unclear. The cytoplasmic tail of gp49A lack tyrosine-based signal transduction motifs. However, it contains a consensus sequence, shared by gp49B, for serine phosphorylation by protein kinase C (CASTELLS et al. 1994). Interestingly, the diversity between gp49A and B, where similar extracellular portions are linked to different types of cytoplasmic tails, is reminiscent of KIR. Members of the KIR family have either long or short cytoplasmic tails, with or without inhibitory tyrosine phosphorylation motifs, (COLONNA and SAMARIDIS 1995; WAGTMANN et al. 1995a). This dual diversity extends to other families of inhibitory receptors, including the type II transmembrane Ly-49 receptors as well as gp49, ILT/LIR/MIR, p91/PIR, and LAIR. The conservation of dual receptor forms in these unrelated receptor families suggests an important role for this feature. Functional gene transfer experiments in NK cells in vitro and in transgenic mice have demonstrated that KIR molecules having long cytoplasmic tails are inhibitory (WAGTMANN et al. 1995b; CAMBIAGGI et al. 1997). In contrast, the function of molecules with short cytoplasmic tails is enigmatic, although some in vitro data have demonstrated that they have activating potential (LANIER 1998). Short-tailed isoforms of KIR, Ly-49, and NKG2 are linked to transmembrane domains containing a charged amino acid. These charged residues permit association with the ITAM-containing DAP12/KARAP homodimers which can generate activating signals (OLCESE et al. 1997; LANIER et al. 1998a, b; SMITH et al. 1998). Similarly, charged residues are present in the transmembrane portion of short-tailed members of the ILT/LIR/MIR, p91/PIR, and LAIR families. In contrast, the gp49A molecule lacks such a charged residue, raising the possibility that it may signal by other mechanisms than via DAP12/KARAP. Thus, further studies will be necessary to determine the signaling capability of gp49A. Identification of the natural ligand for gp49 should facilitate such studies and characterization of the physiological role of gp49 in the regulation of mast and NK cells.

References

Arm JP, Gurish MF, Reynolds DS, Scott HC, Gartner CS, Austen KF, Katz HR (1991) Molecular cloning of gp49, a cell-surface antigen that is preferentially expressed by mouse mast cell progenitors and is a new member of the immunoglobulin superfamily. J Biol Chem 266:15966–15973

Arm JP, Nwankwo C, Austen KF (1997) Molecular identification of a novel family of human Ig superfamily members that possess immunoreceptor tyrosine-based inhibition motifs and homology to the mouse gp49B1 inhibitory receptor. J Immunol 159:2342–2349

Binstadt BA, Brumbaugh KM, Dick CJ, Scharenberg AM, Williams BL, Colonna M, Lanier LL, Kinet J-P, Abraham RT, Leibson PJ (1996) Sequential involvement of Lck and SHP-1 with MHC-recognizing receptors on NK cells inhibits FcR-initiated tyrosine kinase activation. Immunity 5:629–638

Biron CA (1997) Activation and function of natural killer cell responses during viral infections. Curr Opin Immunol 9:24–34

Borges L, Hsu M-L, Fanger N, Kubin M, Cosman D (1997) A family of human lymphoid and myeloid Ig-like receptors, some of which bind to MHC Class I molecules. J Immunol 159:5192–5196

Burshtyn DN, Scharenberg AM, Wagtmann N, Rajagopalan S, Berrada K, Yi T, Kinet J-P, Long EO (1996) Recruitment of tyrosine phosphatase HCP by the killer cell inhibitory receptor. Immunity 4:77–85

Burshtyn DN, Long EO (1997) Regulation through inhibitory receptors: lessons from natural killer cells. Trends Cell Biol 7:473–479

Burshtyn DN, Yang W, Yi T, Long EO (1997) A novel phosphotyrosine motif with a critical amino acid at position −2 for the SH2 domain-mediated activation of the tyrosine phosphatase SHP-1. J Biol Chem 272:13066–13072

Cambiaggi A, Verthuy C, Naquet P, Romagné F, Ferrier P, Biassoni R, Moretta A, Moretta L, Vivier E (1997) NK-cell acceptance of H-2 mismatch bone-marrow grafts in transgenic mice expressing HLA-Cw3 specific killer-cell inhibitory receptor (CD158b). Proc Natl Acad Sci (USA) 94: 8088–8092

Castells MC, Wu X, Arm JP, Austen KF, Katz HR (1994) Cloning of the gp49B gene of the immunoglobulin superfamily and demonstration that one of its two products is an early-expressed mast cell surface protein originally described as gp49. J Biol Chem 269:8393–8401

Colonna M, Samaridis J (1995) Cloning of immunoglobulin-superfamily members associated with HLA-C and HLA-B recognition by human natural killer cells. Science 268:405–408

Cosman D, Fanger N, Borges L, Kubin M, Chin W, Peterson L, Hsu M-L (1997) A novel immunoglobulin superfamily receptor for cellular and viral MHC class I molecule. Immunity 7:273–282

Daëron M, Malbec O, Latour S, Arock M, Fridman WH (1995) Regulation of high-affinity IgE receptor-mediated mast cell activation by murine low-affinity IgG receptors. J Clin Invest 95:577–585

Daëron M (1997) Fc receptor biology. Annu Rev Immunol 15:203–234

Echtenacher B, Mannel DN, Hultner L (1996) Critical protective role of mast cells in a model of acute septic peritonitis. Nature 381:75–77

Flemming TJ, Donnadieu E, Song CH, Van Laethem F, Galli SJ, Kinet J-P (1997) Negative regulation of FcεRI-mediated degranulation by CD81. J Exp Med 186:1307–1314

Gupta N, Scharenberg AM, Burshtyn DN, Wagtmann N, Lioubin MN, Rohrschneider LR, Kinet J-P, Long EO (1997) Negative signaling pathways of the killer cell inhibitory receptor and FcγRIIb1 require distinct phosphatases. J Exp Med 186:473–478

Guthmann MD, Tal M, Pecht I (1995) A secretion inhibitory signal transduction molecule on mast cells is another C-type lectin. Proc Natl Acad Sci USA 92:9397–9401

Hayami K, Fukuta D, Nishikawa Y, Yamashita Y, Inui M, Ohyama Y, Hikida M, Ohmori H, Takai T (1997) Molecular cloning of a novel murine cell-surface receptor glycoprotein homologous to killer cell inhibitory receptors. J Biol Chem 272:7320–7327

Katz HR, Vivier E, Castells MC, McCormick MJ, Chambers JM, Austen KF (1996) Mouse mast cell gp49B1 contains two immunoreceptor tyrosine-based inhibition motifs and suppresses mast cell activation upon ligation with FcεRI. Proc Natl Acad Sci (USA) 93:10809–10814

Kubagawa H, Burrows PD, Cooper MD (1997) A novel pair of immunoglobulin-like receptors expressed by B-cells and myeloid cells. Proc Natl Acad Sci (USA) 94:5261–5266

Kuroiwa A, Yamashita Y, Inui M, Yuasa T, Ono M, Nagabukuro A, Matsuda Y, Takai T (1998) Association of tyrosine phosphatase SHP-1 and SHP-2, inositol 5-phosphatase SHIP with gp49B1, and chromosomal assignment of the gene. J Biol Chem 273:1070–1074

Lanier LL (1998) Natural killer cell receptors. Ann Rev Immunol 16:359–393

Lanier LL, Corliss B, Wu J, Phillips JH (1998a) Association of DAP-12 with activating CD94/NKG2 C natural killer cell receptors. Immunity 8:693–701

Lanier LL, Corliss BC, Wu J, Leong C, Phillips JH (1998b) Immunoreceptor DAP12 bearing a tyrosine-based activation motif is involved in activating natural killer cells. Nature 391:703–707

LeBlanc PA, Biron CA (1984) Mononuclear phagocyte maturation: a cytotoxic monoclonal antibody reactive with postmonoblast stages. Cell Immunol 83:242–254

Ljunggren H-G, Kärre K (1990) In search of the "missing self": MHC molecules and natural killer cell recognition. Immunol Today 11:237

Malaviya R, Ikeda T, Ross E, Abraham SN (1996) Mast cell modulation of neutrophil influx and bacterial clearance at sites of infection through TNF-alpha. Nature 381:77–80

Meyaard L, Adema GJ, Chang C, Woollatt E, Sutherland E, Lanier LL, Phillips JH (1997) LAIR-1, a novel inhibitory receptor expressed on human mononuclear leukocytes. Immunity 7:283–290

Olcese L, Cambiaggi A, Semenzato G, Bottino C, Moretta A, Vivier E (1997) Killer-cell activatory receptors for MHC Class I molecules are included in a multimeric complex expressed by human killer cells. J Immunol 158:5083–5086

Ono M, Okada H, Bolland S, Yanagi S, Kurosaki T, Ravetch JV (1997) Deletion of SHIP or SHP-1 reveals two distincts pathways for inhibitory signaling. Cell 90:293–301

Rojo S, Burshtyn DN, Long EO, Wagtmann N (1997) Type I transmembrane receptor with inhibitory function in mouse mast cells and natural killer cells. J Immunol 158:9–12

Samaridis J, Colonna M (1997) Cloning of novel immunoglobulin superfamily receptors expressed on human myeloid and lymphoid cells: structural evidence for new stimulatory and inhibitory pathways. Eur J Immunol 27:660–665

Smith KM, Wu J, Bakker ABH, Phillips JH, Lanier LL (1998) Ly-49D and Ly-49H associate with mouse DAP12 and form activating receptors. J Immunol 161:7–10

Trinchieri G (1995) Natural killer cells wear different hats: effector cells of innate resistance and regulatory cells of adaptive immunity and of hematopoiesis. Semin Immunol 7:83–88

Wagtmann N, Biassoni R, Cantoni C, Verdiani S, Malnati M, Vitale M, Bottino C, Moretta L, Moretta A, Long EO (1995a) Molecular clones of p58 natural killer cell receptor reveal immunoglobulin-related molecules with diversity in both the extra- and the intracellular domains. Immunity 2:439–449

Wagtmann N, Rajogopalan S, Winter CC, Peruzzi M, Long EO (1995b) Killer cell inhibitory receptors specific for HLA-C and HLA-B identified by direct binding and by functional transfer. Immunity 3:801–809

Wagtmann N, Rojo S, Eichler E, Mohrenweiser H, Long EO (1997) A new human gene complex encoding the killer cell inhibitory receptors and related monocyte/macrophage receptors. Curr Biol 7:615–618

Wang LL, Mehta IK, LeBlanc PA, Yokoyama WM (1997) Mouse natural killer cells express gp49B1, a structural homologue of human killer inhibitory receptors. J Immunol 158:13–17

Winter CC, Long EO (1997) A single amino acid in the p58 killer cell inhibitory receptor controls the ability of natural killer cells to discriminate between the two groups of HLA-C allotypes. J Immunol 158:4026–28

A Novel Family of Inhibitory Receptors for HLA Class I Molecules That Modulate Function of Lymphoid and Myeloid Cells

M. Colonna[1], F. Navarro[2], and M. López-Botet[2]

1 The Immunoglobulin-Like Transcripts Receptor Family

Immunoglobulin-like transcripts (ILTs) encode several Ig-SF receptors which are structurally and functionally related to killer cell inhibitory receptors (KIRs) and are expressed on lymphoid and/or on myeloid cells (Yokoyama 1997; Table 1). ILTs are characterized by two or four homologous extracellular Ig-SF domains and can be classified by differing transmembrane and cytoplasmic domains (Samaridis and Colonna 1997). One subset of ILT receptors displays long cytoplasmic tails containing immunoreceptor tyrosine-based inhibitory motifs (ITIMs). These receptors mediate inhibition of cell activation by recruiting protein tyrosine phosphatase SHP-1 (Cella et al. 1997a; Cosman et al. 1997; Colonna et al. 1997; Arm et al. 1997; Colonna et al. 1998). Another subset of ILT receptors contains short cytoplasmic domains that lack kinase homology or recognizable motifs for signaling mediators. In addition, they are characterized by the presence of a single basic arginine residue within the hydrophobic transmembrane domain (Colonna et al. 1997; Borges et al. 1997). These ILT receptors closely resemble activating natural killer (NK) cell receptors, which share a positively charged lysine residue in

[1] Basel Institute for Immunology, 487 Grenzacherstrasse, CH-4005 Basel, Switzerland
[2] Servicio de Inmunologia, Hospital Universitario de la Princesa, Diego de Leon 62, 28006 Madrid, Spain

Table 1. Human ILT/LIR/MIR Receptors

Receptor					Ligand
Function	Alternative names	Structure	MAb	mw	
Inhibitory					
ILT2	LIR1, MIR7	4 Ig-SF domains	HP-F1	110	HLA-A, -B and -G1
ILT3	LIR5	2 Ig-SF domains	ZM3.8, ZM4	60	unknown
ILT4	LIR2, MIR10	4 Ig-SF domains	42D1	110	HLA-A, -B and -G1
ILT5	LIR3, HL9	4 Ig-SF domains	–	unknown	unknown
LIR8		4 Ig-SF domains		unknown	unknown
Activating					
ILT1	LIR7	4 Ig-SF domains	135	60	unknown
ILT1-like protein		4 Ig-SF domains	–	unknown	unknown
LIR6a		4 Ig-SF domains	–	unknown	unknown
Soluble					
ILT6	LIR4, HM31, HM43	4 Ig-SF domains	–	unknown	unknown

the transmembrane domain and a short cytoplasmic domain that lacks sequence motifs implicated in signal transduction (BIASSONI et al. 1996). To transduce signals, activating NK cell receptors associate with an immunoreceptor tyrosine-based activation motif (ITAM)-containing subunit called DAP12 (OLCESE et al. 1997; LANIER et al. 1998a, b; SMITH et al. 1998; CAMPBELL et al. 1998). Because of these structural similarities, it seems most likely that ILT receptors with short cytoplasmic tails also activate cells and use an associated protein to transduce stimulatory signals. A third subset of ILTs has no transmembrane and cytoplasmic domains and may be secreted as soluble receptors (COLONNA et al. 1997; ARM et al. 1997; BORGES et al. 1997). All the ILT genes map on human chromosome 19q13.4 in close linkage with the KIR genes (WAGTMANN et al. 1997; TORKAR et al. 1998; WENDE et al. Mammalian Genome, in press). Their murine counterparts have been recently identified (HAYAMI et al. 1997; KUBAGAWA et al. 1997).

Here we will review cellular distribution, biochemical characteristics and function of the ITIM-containing ILTS which includes ILT2, ILT3, ILT4 and ILT5.

2 Cellular Distribution of ITIM-Containing ILTs

ITIM-bearing ILT receptors are expressed on lymphoid and/or on myeloid cells. The expression patterns of many ILTs have been determined by cell surface staining with specific monoclonal antibodies (mAbs). In peripheral blood leukocytes, ILT2 is expressed on subsets of NK and T cells, and on all B lymphocytes and monocytes

(COLONNA et al. 1997). In the bone marrow, ILT2 is not expressed in pre-B cells, but expression increases with B cell maturation. In peripheral lymphoid tissues, ILT2 is expressed on germinal center and plasma cells (COLONNA, unpublished data). ILT2 is also expressed on liver sinusoidal macrophages (LOPEZ-BOTET, unpublished data), and on macrophages and dendritic cells (DCs) cultured from monocytes using appropriate cytokines (COLONNA et al. 1997). In contrast to ILT2, other ILT receptors show a more restricted pattern of expression. ILT3 and ILT4 are expressed on monocytes, macrophages derived from purified monocytes, immature DCs, and DCs stimulated either with bacterial products, inflammatory cytokines, or via CD40-CD40L interactions to induce maturation (CELLA et al. 1997a; COLONNA et al. 1998). Cell surface expression of ILT3 and ILT4 has not been detected on NK cells, T cells, EBV-transformed B cell lines, peripheral B cells, or neutrophils (CELLA et al. 1997a; COLONNA et al. 1998), although mRNA has been detected by RT-PCR studies in lymphocytes (BORGES et al. 1997). This mRNA may not be correctly spliced or translated, and hence the protein is not expressed on the cell surface. Alternatively, RT-PCR experiments may detect yet unknown mRNAs which cross-hybridize with ILT3 or ILT4.

ILT5 has been exclusively cloned from RNA derived from myelomonocytic cells and may have an expression pattern similar to that of ILT3 and ILT4 (COLONNA et al. 1997; ARM et al. 1997; BORGES et al. 1997). Interestingly, analysis of ILT structural variability within the population has revealed that ILT5 is strikingly diverse (COLONNA et al. 1997). Amplification and sequencing of ILT5 cDNA from a pool of cells derived from different individuals has shown 15 distinct cDNAs, whereas only one ILT5 cDNA clone was found from a single donor. Thus, ILT5 may be extensively polymorphic, with amino acid variants clustered in several distinct regions of the extracellular domains. Genomic studies are necessary to establish whether all of the cloned ILT5 variants are encoded by different alleles of the same locus, or by different loci.

3 Specificity of ITIM-Bearing ILT Receptors

The homology between ILTs and KIRs suggests that ILTs may be receptors for HLA class I molecules. Indeed, ILT2 and ILT4 have been shown to interact with HLA class I molecules with a broad specifity (COLONNA et al. 1997, 1998; COSMAN et al. 1997). Soluble ILT2 and ILT4 bind to all the tested HLA-A and -B transfectants made in the class I-negative cell line 721.221, but not to HLA-C transfectants. Interaction between ILT2, ILT4 and class I molecules has also been confirmed by studying the binding of soluble class I molecules to ILT2- and ILT4-transfected cells. Soluble HLA-A*0201, -B*0801, -B*2702, and -B*3501 tetramers complexed either with influenza- or HIV-derived viral peptides bind to ILT2 and ILT4-transfected COS cells (COLONNA et al. 1998; Colonna, unpublished data), but not to untransfected COS cells.

In addition to HLA-A and -B, ILT2 and ILT4 interact with the non-classical class I molecule HLA-G1 (COLONNA et al. 1997, 1998). This interaction is noteworthy, since HLA-G is selectively expressed in the trophoblast, a tissue of fetal origin devoid of HLA-A, and -B molecules that separate the developing embryo from the mother. Thus, interaction of ILT2 and ILT4 with HLA-G may inhibit decidual leukocytes, contributing to maternal tolerance of the fetal semiallograft. Interestingly, soluble ILT2/LIR1 has been also shown to bind to UL18, a class I-like molecule synthesized by the human cytomegalovirus (HCMV) (COSMAN et al. 1997). UL18 may be expressed by HCMV-infected cells in an attempt to engage leukocyte inhibitory receptors and block cell-mediated anti-viral responses. However, it has been difficult to detect cell surface expression of UL18 upon transfection of the UL18 gene (REYBURN et al. 1997; Leong et al. 1998) or during HCMV infection. Thus, the functional significance of ILT2 interaction with UL18 is still unclear and should be investigated during the actual infection of a cell or an individual.

It has been shown that CD94/NKG2 receptors recognize a non-classical class I molecule, called HLA-E (BRAUD et al. 1998; BORREGO et al. 1998; LEE et al. 1998b; CARRETERO et al. 1998). One interesting feature of HLA-E is that its cell surface expression is stabilized by a nonamer derived from the leader sequence of many HLA-A, -B, and -C molecules as well as HLA-G1 (BRAUD et al. 1997; LEE et al. 1998a). As a consequence, transfection of many HLA class I genes in the class I-negative 721.221 cells results in the coexpression of HLA-E. This raises the question of whether ILT2 and ILT4 interact with HLA-A, HLA-B, and HLA-G1 molecules on 721.221 transfectants, or, like CD94/NKG2 receptors, with the coexpressed HLA-E molecule. To address this problem, we mutated the HLA-G1 leader sequence to encode a peptide that does not stabilize HLA-E and used it to generate a 721.221-G1 transfectant that does not co-express HLA-E. We then tested the binding of soluble ILT2 to the 721.221-HLA-G + /E-, HLA-G + /E + , and HLA-E + transfectants and found that ILT2 strongly interacts with HLA-G1, whereas there is no detectable interaction with HLA-E (NAVARRO et al. Eur J Immunol, in press).

In contrast to ILT2 and ILT4, other ILTs did not appear to bind MHC class I molecules. Soluble ILT3 and ILT5 proteins do not bind to class I transfectants and none of the tested class I tetramers bound ILT5-transfected COS cells (CELLA et al. 1997a; COLONNA, unpublished data). No information is yet available on ILT3 or ILT5 ligands, but it is possible that ILT3 and ILT5 are receptors for MHC class I-related molecules, such as CD1 (PORCELLI 1995), MR1 (HASHIMOTO et al. 1995) and MIC (BAHRAM et al. 1994).

4 ITIM-Bearing ILT Receptors Deliver a Negative Signal That Inhibits Functional Responses and Early Biochemical Events

To determine whether ITIM-bearing ILTs can inhibit cell activation, ILTs have been stably expressed in rat basophilic leukemia (RBL) cells, which release serotonin upon engagement of the Fc receptor for IgE (FcεRI) (COLONNA et al. 1997, 1998). Secretion of serotonin triggered via the FcεRI was inhibited when ILT2, ILT4, or ILT5 were ligated to FcεRI, using anti-ILT mAbs and IgE immobilized on plastic to mimic ILT and FcεRI ligands, respectively. Inhibition of serotonin secretion was also obtained by incubating ILT4-transfected RBL cells with class I-transfected 721.221 cells coated with TNP and mouse anti-TNP IgE. In contrast, no inhibition was observed by incubating ILT5-transfected RBL cells with TNP-IgE-coated class I transfectants, confirming that interaction of ILT4 (but not ILT5) with class I molecules inhibits functional responses in RBL cells.

The physiological responses controlled by ILT receptors in leukocytes are not completely understood. There is extensive evidence that ILT2 mediates inhibition of NK cell-mediated cytotoxicity (COLONNA et al. 1997). In addition, ILT2 increases the activation threshold of CD8 + T cells (COLONNA et al. 1997); consequently, high antigenic doses are required to activate ILT2-expressing T cells. ILTs are also capable of an inhibitory function in myelomonocytic and B cells when recruited to a stimulatory receptor. For example, co-engagement of ILTs in B cells inhibits Ca^{2+} mobilization triggered via the B cell receptor (BCR) (COLONNA et al. 1997). Similarly, coligation of ILTs in monocytes inhibits Ca^{2+} mobilization triggered via the Fc γ receptor II (FcγRII) and HLA-DR (CELLA et al. 1997; COLONNA et al. 1997, 1998). However, the functional responses controlled in B cells and monocytes are still unclear. In B cells, ILTs may modulate the threshold of antigen-mediated B cell activation. In myelomonocytic cells, ILTs may control antigen-presenting functions, such as antigen uptake and presentation, migratory capacity, cytokine production, and costimulatory function (CELLA et al. 1997b). ILTs may also control inflammatory responses mediated by monocyte-macrophages, such as oxidative burst, or inhibit their cytotoxicity against normal class I-expressing cells, hence allowing lysis of tumor cells that have lost expression of self-class I molecules (PHILIP and EPSTEIN 1986).

5 ITIM-Bearing ILT Receptors Associate with SHP-1 Phosphatase

It has been shown that negative signaling through KIRs is mediated by recruitment of SHP-1 and SHP-2 phosphatases upon tyrosine phosphorylation of cytoplasmic ITIMs (BURSHTYN et al. 1996; CAMPBELL et al. 1996; OLCESE et al. 1996; FRY et al. 1996; BINSTADT et al. 1996). The cytoplasmic tails of ILT2, 3, 4, and 5 contain three to four tyrosine-based motifs similar to those found in KIRs, although only some fit the V/I-x-Y-x-x-L motif proposed to be required for SHP-1 binding (BURSHTYN et al. 1997). To determine whether ILTs also recruit SHP phosphatases, ILT2, ILT3, and ILT4 were immunoprecipitated from NK cells, B cells, and monocytes either resting or stimulated with pervanadate, which induces substantial tyrosine phosphorylation of cellular substrates.

Immunoprecipitates were immunoblotted with anti-SHP-1, anti-SHP-2, and anti-SHIP antibodies, showing that all ITIM-bearing ILTs are associated with SHP-1 after treatment with pervanadate, whereas there is no detectable association with SHP-2 and SHIP (CELLA et al. 1997; COSMAN et al. 1997; COLONNA et al. 1997, 1998).

6 Concluding Remarks

A novel family of Ig-like receptors expressed on lymphoid and myeloid cells has recently been discovered. Some of these receptors mediate an inhibitory function upon specific recognition of HLA class I molecules. Rapid progress can be foreseen in the characterization of these inhibitory receptors with regard to their biological role. Our working hypothesis, based on current information available on the more extensively characterized KIRs and ITIM-bearing receptors in B cells (DAERON et al. 1995; ONO et al. 1996), is that ILTs establish an activation threshold that must be overcome by triggering stimuli in order to promote cellular activation efficiently. A number of questions remain unanswered about ILT ligand specificities, binding affinities, and structure: (a) Despite being broadly reactive with HLA-A, -B, and -G molecules, why doesn't ILT2 efficiently interact with all HLA class I allotypes? (b) What are the ligand(s) of ILTs that apparently do not interact with HLA class I molecules? (c) What is the significance of the allelic polymorphism found in some ILTs (i.e. ILT5)? (d) How is the selective expression of ILT2 on T lymphocytes and NK cell subsets regulated?

Finally, it will be important to see whether ILT-class I interactions are important in diseases. Viruses may escape from immune responses by producing glycoproteins, like UL18, which interact with ILT2, potentially blocking lymphoid and myeloid cell activation. On the other hand, a dysfunction of the ILT inhibitory machinery may facilitate inappropriate activation of immune cells, contributing to the pathogenesis of some chronic inflammatory autoimmune diseases.

Acknowledgments. We thank Marina Cella and Susan Gilfillan (Basel Institute for Immunology) for reviewing the manuscript. The Basel Institute for Immunology was founded and is supported by Hoffmann-La Roche Ltd, CH-4002 Basel. The work at Hospital de la Princesa is supported by the grants SAF98-0006 (Plan Nacional I + D) and 08/005/97 (C.A.M.).

References

Arm JP, Nwankwo C, Austen KF (1997) Molecular identification of a novel family of human Ig superfamily members that possess immunoreceptor tyrosine-based inhibition motifs and homology to the mouse gp49B1 inhibitory receptor. J Immunol 159:2342–2349
Bahram S, Bresnahan M, Geraghty DE, Spies T (1994) A second lineage of mammalian major histocompatibility complex class I genes. Proc Natl Acad Sci USA 91:6259–6263

Biassoni R, Cantoni C, Falco M, Verdiani S, Bottino C, Vitale M, Conte R, Poggi A, Moretta A, Moretta L (1996) The human leukocyte antigen (HLA)-C-specific "activatory" or "inhibitory" natural killer cell receptors display highly homologous extracellular domains but differ in their transmembrane and intracytoplasmic portions. J Exp Med 183:645–650

Binstadt BA, Brumbaugh KM, Dick CJ, Scharenberg AM, Williams BL, Colonna M, Lanier LL, Kinet J-P, Abraham RT, Leibson, PJ (1996) Sequential Involvement of Lck and SHP-1 with MHC-Recognizing Receptors on NK Cells Inhibits FcR-Initiated Tyrosine Kinase Activation. Immunity 5:629–638

Borges L, Hsu ML, Fanger N, Kubin M, Cosman D (1997) A family of human lymphoid and myeloid Ig-like receptors, some of which bind to MHC class I molecules. J Immunol 159:5192–5196

Borrego F, Ulbrecht M, Weiss EH, Coligan JE, Brooks AG (1998) Recognition of human histocompatibility leukocyte antigen (HLA)-E complexed with HLA class I signal sequence-derived peptides by CD94/NKG2 confers protection from natural killer cell-mediated lysis. J Exp Med 187:813–818

Braud V, Jones EY, McMichael A (1997) The human major histocompatibility complex class Ib molecule HLA-E binds signal sequence-derived peptides with primary anchor residues at positions 2 and 9. Eur J Immunol 27:1164–1169

Braud VM, Allan DS, O'Callaghan CA, Soderstrom K, D'Andrea A, Ogg GS, Lazetic S, Young NT, Bell JI, Phillips JH, Lanier LL, McMichael AJ (1998) HLA-E binds to natural killer cell receptors CD94/NKG2 A, B, and C. Nature 391:795–799

Burshtyn DN, Scharenberg AM, Wagtmann N, Rajagopalan S, Berrada K, Yi T, Kinet JP, Long EO (1996) Recruitment of tyrosine phosphatase HCP by the killer cell inhibitor receptor. Immunity 4:77–85

Burshtyn DN, Yang W, Yi T, Long EO (1997) A novel phosphotyrosine motif with a critical amino acid at position –2 for the SH2 domain-mediated activation of the tyrosine phosphatase SHP-1. J Biol Chem 16, 272:13066–13072

Campbell KS, Dessing M, Lopez Botet M, Cella M, Colonna M (1996) Tyrosine phosphorylation of a human killer inhibitory receptor recruits protein tyrosine phosphatase 1 C. J Exp Med 184:93–100

Campbell KS, Cella M, Carretero M, Lopez-Botet M, Colonna M (1998) Signaling through human killer cell activating receptors triggers tyrosine phosphorylation of an associated protein complex. Eur J Immunol 28:599–609

Carretero M, Palmieri G, Llano M, Tullio V, Santoni A, Geraghty DE, Lopez-Botet M (1998) Specific engagement of the CD94/NKG2-A killer inhibitory receptor by the HLA-E class Ib molecule induces SHP-1 phosphatase recruitment to tyrosine -phosphorylated NKG2-A: evidence for receptor function in heterologous transfectants. Eur J Immunol 28:280–291

Cella M, Dohring C, Samaridis J, Dessing M, Brockhaus M, Lanzavecchia A, Colonna M (1997a) A novel inhibitory receptor (ILT3) expressed on monocytes, macrophages, and dendritic cells involved in antigen processing. J Exp Med 185:1743–1751

Cella M, Sallusto F, Lanzavecchia A (1997b) Origin, maturation and antigen presenting function of dendritic cells. Curr Opin Immunol 9:10–16

Colonna M, Navarro F, Bellon T, Llano M, Garcia P, Samaridis J, Angman L, Cella M, Lopez-Botet M (1997) A common inhibitory receptor for major histocompatibility complex class I molecules on human lymphoid and myelomonocytic cells. J Exp Med 186:1809–1818

Colonna M, Samaridis J, Cella M, Angman L, Allen R, O'Callaghan C, Dunbar R, Ogg G, Cerundolo V, Rolink A (1998) Human myelomonocytic cells express an inhibitory receptor for classical and non-classical MHC class I molecules. J Immunol 160:3096–3100

Cosman D, Fanger N, Borges L, Kubin M, Chin W, Peterson L, Hsu ML (1997) A novel immunoglobulin superfamily receptor for cellular and viral MHC class I molecules. Immunity 7:273–282

Daeron M, Latour S, Malbec O, Espinosa E, Pina P, Pasmans S, Fridman WH (1995) The same tyrosine-based inhibition motif, in the intracytoplasmic domain of Fc gamma RIIB, regulates negatively BCR-, TCR-, and FcR-dependent cell activation. Immunity 3:635–646

Fry AM, Lanier LL, Weiss A (1996) Phosphotyrosines in the killer cell inhibitory receptor motif of NKB1 are required for negative signaling and for association with protein tyrosine phosphatase 1 C. J Exp Med 184:295–300

Hayami K, Fukuta D, Nishikawa Y, Yamashita Y, Inui M, Ohyama Y, Hikida M, Ohmori H, Takai T (1997) Molecular cloning of a novel murine cell-surface glycoprotein homologous to killer cell inhibitory receptors. J Biol Chem 272:7320–7327

Hashimoto K, Hirai M, Kurosawa Y (1995) A gene outside the human MHC related to classical HLA class I genes. Science 269:693–695

Kubagawa H, Burrows PD, Cooper MD (1997) A novel pair of immunoglobulin-like receptors expressed by B cells and myeloid cells. Proc Natl Acad Sci USA 94:5261–5266

Lanier LL, Corliss B, Wu J, Leong C, Phillips JH (1998a) Immunoreceptor DAP12 bearing a tyrosine-based activation motif is involved in activating NK cells. Nature 391:703–707

Lanier LL, Corliss B, Wu J, Phillips JH (1998b) Association of DAP12 with activating CD94/NKG2 C NK cell receptors. Immunity 8:693–701

Lee N, Goodlett DR, Ishitani A, Marquardt H, Geraghty, DE (1998a) HLA-E surface expression depends on binding of TAP-dependent peptides derived from certain HLA class I signal sequences. J Immunol 160:4951–4960

Lee N, Llano M, Carretero M, Ishitani A, Navarro F, Lopez-Botet M, Geraghty DE (1998b) HLA-E is a major ligand for the natural killer inhibitory receptor CD94/NKG2 A. Proc Natl Acad Sci USA 95:5199–5204

Leong CC, Chapman TL, Bjorkman PJ, Formankova D, Mocarski ES, Phillips JH, Lanier LL (1998) Modulation of natural killer cell cytotoxicity in human cytomegalovirus infection: the role of endogenous class I major histocompatibility complex and a viral class I homolog. J Exp Med 187:1681–1687

Olcese L, Lang P, Vely F, Cambiaggi A, Marguet D, Blery M, Hippen KL, Biassoni R, Moretta A, Moretta L, Cambier JC, Vivier E (1996) Human and mouse killer-cell inhibitory receptors recruit PTP1 C and PTP1D protein tyrosine phosphatases. J Immunol 156:4531–4534

Olcese L, Cambiaggi A, Semenzato G, Bottino C, Moretta A, Vivier EJ (1997) Human killer cell activatory receptors for MHC class I molecules are included in a multimeric complex expressed by natural killer cells. J Immunol 158:5083–5086

Ono M, Bolland S, Tempst P, Ravetch JV (1996) Role of the inositol phosphatase SHIP in negative regulation of the immune system by the receptor Fc(gamma)RIIB. Nature 383:263–266

Philip R, Epstein LB (1986) Tumour necrosis factor as immunomodulator and mediator of monocyte cytotoxicity induced by itself, gamma-interferon, and interleukin-1. Nature 323:86–89

Porcelli SA (1995) The CD1 family: a third lineage of antigen-presenting molecules. Adv Immunol 59:1–98

Reyburn HT, Mandelboim O, Vales-Gomez M, Davis DM, Pazmany L, Strominger JL (1997) The class I MHC homologue of human cytomegalovirus inhibits attack by natural killer cells. Nature 386:514–517

Samaridis J, Colonna M (1997) Cloning of novel immunoglobulin superfamily receptors expressed on human myeloid and lymphoid cells: structural evidence for new stimulatory and inhibitory pathways. Eur J Immunol 27:660–665

Smith KM, Wu J, Bakker AB, Phillips JH, Lanier LL (1998) Ly-49D and Ly-49H associate with mouse DAP12 and form activating receptors. J Immunol 161:7–10

Torkar M, Norgate Z, Colonna M, Trowsdale J, Wilson MJ (1998) Isotypic variation in arrangement of Immunoglobulin-like transcripts/killer cell inhibitory receptor loci in the leukocyte receptor complex. Eur J Immunol 28:3959–3967

Wagtmann N, Rojo S, Eichler E, Mohrenweiser H, Long EO (1997) A new human gene complex encoding the killer cell inhibitory receptors and related monocyte/macrophage receptors. Curr Biol 7:615–618

Yokoyama WM (1997) What goes up must come down: the emerging spectrum of inhibitory receptors. J Exp Med 186:1803–1808

Interactions of LIRs, a Family of Immunoreceptors Expressed in Myeloid and Lymphoid Cells, with Viral and Cellular MHC Class I Antigens

L. Borges, N. Fanger, and D. Cosman

1 Introduction

Human cytomegalovirus (hCMV) infects 70%–90% of adults (Wentworth and Alexander 1971). Evidence of CMV infection, but not other common viruses such as measles or influenza, has been found in the remote Tiriyo Indian tribe in Brazil, suggesting that CMV has been present in the human population for a long time (Gold and Nankervis 1976). hCMV infection usually occurs early in life with only mild and subclinical disease, followed by virus latency that can last for the lifetime of immunocompetent individuals. However, reactivation of the virus can occur in immuno-compromised patients, such as AIDS and transplant patients, and become a major cause of morbidity and mortality.

hCMV has evolved several different strategies to escape immunosurveillance and modulate the host immune response. For instance, expression of the viral protein pp65 blocks the proteolysis of viral proteins in the cytoplasm of infected cells and prevents the generation of viral peptides that could be presented to cytotoxic T cells (Gilbert et al. 1996). The US6 protein interferes with the peptide transporter TAP and blocks peptide delivery to MHC class I molecules (Hengel et al. 1997). Other mechanisms are used to inhibit cell surface expression of the host

Immunex Corporation, 51 University Street, Seattle, WA 98101-2936, USA

MHC class I molecules; the US3 viral protein binds to class I molecules and retains them in the ER (AHN et al. 1996); the US2 (WIERTZ et al. 1996a) and US11 (WIERTZ et al. 1996b) proteins interact with newly synthesized class I molecules and prevent them from entering the ER. Even though the down-regulation of MHC class I molecules could limit T cell cytotoxicity towards CMV-infected cells, this strategy would, in theory, expose the cells to cytotoxic responses mediated by NK cells. These cells express killer inhibitory receptors (KIRs), which recognize MHC class I molecules and send inhibitory signals that prevent NK cytotoxicity. In vitro studies have shown that blockade of the KIR-class I interaction releases the NK cells from inhibition and triggers cytotoxicity. In vivo, several lines of evidence indicate that NK cells play a crucial role in controlling CMV infections in mice and men. Depletion of NK cells in mice leads to an increase in susceptibility to CMV infection (BUKOWSKI et al. 1984). In humans, the most compelling evidence for the role of NK cells in limiting CMV infections comes from a teenage girl who completely lacked NK cells, but otherwise displayed normal T and B cell responses, and had an unusual sensitivity to recurrent infections with CMV and other herpes viruses (BIRON et al. 1989).

hCMV and murine CMV (mCMV) encode molecules that are homologous to MHC class I antigens. These molecules include the hCMV protein UL18 and the mCMV protein m144. Similar to MHC class I molecules, both UL18 and m144 associate with β2-microglobulin (BROWNE et al. 1990; CHAPMAN and BJORKMAN 1998). In addition, UL18, but not m144, binds to endogenous peptides in order to form a thermally stable complex (CHAPMAN and BJORKMAN 1998). m144 has a major deletion in the putative peptide binding groove, probably precluding the molecule from associating with peptides. It has been proposed that UL18 and m144 could act as class I decoys to engage KIRs and modulate NK cytotoxicity towards CMV-infected cells that have down-regulated expression of the host MHC class I (FAHNESTOCK et al. 1995). Deletion of m144 from the mCMV genome does not affect viral replication in vitro, but considerably decreases virulence in vivo (FARRELL et al. 1997). Virulence of the deletion mutant can be restored if NK cells are depleted (FARRELL et al. 1997). The strict species specificity of hCMV precludes similar in vivo testing of UL18 deletion mutants. However, we have taken a different approach to define the molecular and cellular targets of UL18.

2 The Leukocyte Immunoglobulin-like Receptor Family

To identify cellular counterstructures for UL18, we constructed a fusion protein of the extracellular domain of UL18 and the Fc region of human IgG1. Flow cytometric analysis using this protein showed substantial binding to both monocytic and B cell lines. After screening a CB-23 expression library with UL18-Fc, we identified a novel immunoglobulin superfamily receptor for UL18 that we named leukocyte immunoglobulin-like receptor-1 (LIR-1) (COSMAN et al. 1997). This molecule is a type 1 transmembrane glycoprotein, composed of four immuno-globulin-like domains in the extracellular region, a single transmembrane domain, and an intracellular region containing four immunoreceptor tyrosine-based inhibitory motif (ITIM)-like sequences. Southern and Northern blot analysis using a LIR-1 cDNA probe soon revealed that LIR-1 was part of a multigene family, expressed both in myeloid and lymphoid cells. This observation was later confirmed with the cloning of additional LIR cDNAs from human PBMC and dendritic cell cDNA libraries (BORGES et al. 1997). Eight different members of this family have now been identified (Fig. 1). They can be subdivided into three groups: transmembrane molecules with two to four ITIM-like sequences (LIR-1, -2, -3, -5, and -8); transmembrane molecules with short cytoplasmic domains, a positively charged arginine residue within the transmembrane domain, and no ITIMs (LIR-6a, -6b, and -7); and soluble molecules with no transmembrane domain (LIR-4). Sequence alignments and comparisons show that these molecules have amino acid sequence identities ranging from 63% to 84% relative to the LIR-1 extracellular domain sequence.

The LIR molecules are closely related to other Ig-superfamily immunoreceptors including the bovine neutrophil FcγR for IgG2 (ZHANG et al. 1995), murine gp49B1/B2 antigens (CASTELLS et al. 1994; ROJO et al. 1997), murine PIR-A and PIR-B/p91 receptors (HAYAMI et al. 1997; KUBAGAWA et al. 1997), the human FcαR (MALISZEWSKI et al. 1990), and the human KIRs (COLONNA 1996; LONG et al. 1996). The level of sequence identity ranges from 39% (human FcαR) to 54% (bovine FγcR for IgG2) when these sequences are compared with the extracellular domain sequence of LIR-1. Other groups, using different strategies, have reported the cloning of the same or closely related cDNAs. Different names are currently being used to describe the same molecules, as shown in Table 1. To avoid further complications, we will use only the LIR nomenclature when referring to our own studies.

The LIR gene family maps to human chromosome 19q13.2-q13.4 (BORGES et al. 1997; WAGTMANN et al. 1997), in close proximity to the cytogenetic loci for the KIR gene complex (SUTO et al. 1996) and FcαR gene (de WIT et al. 1995). Recently, two new immunoglobulin superfamily receptors, PIR-A and PIR-B, were identified in mice (KUBAGAWA et al. 1997) and mapped to chromosome 2, band A2, a region that is syntenic with human chromosome 19q13.2-13.4. The genetic linkage between murine PIRs and human LIRs and KIRs suggests that these gene families have evolved from a common gene ancestor. PIR-A and PIR-B contain six immunoglobulin repeats in the extracellular region, but have very different cyto-plasmic tails. PIR-B has a long cytoplasmic region containing four ITIMs; PIR-A contains a short cytoplasmic tail devoid of ITIMs and has a charged arginine residue in the transmembrane region, a situation that is reminiscent of the type of receptors found within the LIR and KIR gene families. ITIM-containing KIRs deliver inhibitory signals that block NK cell activation, whereas the ITIM-lacking KIRs trigger stimulatory signaling pathways (COLONNA 1996). As will be described later in this review, we have evidence that ITIM-containing LIRs such as LIR-1 and LIR-2 deliver inhibitory signals. In view of the close genetic linkage and se-

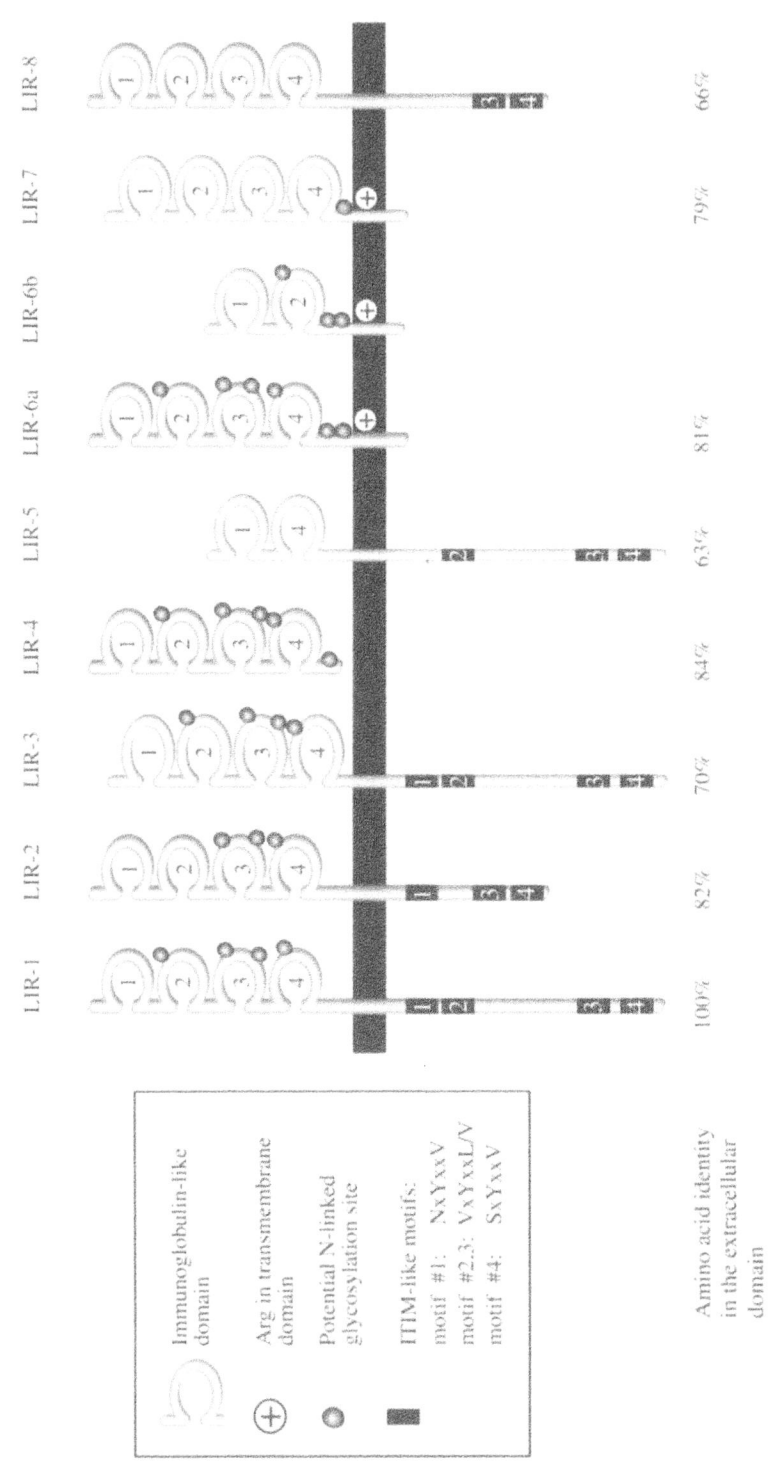

Fig. 1. Schematic representation of the structures of the leukocyte immunoglobulin-like receptor (*LIR*) molecules. The immunoreceptor tyrosine-based inhibitory motif (*ITIM*)-like motifs are numbered 1 through 4 and their consensus sequences are indicated in the legend. LIR-6a and LIR-6b are encoded by alternatively spliced mRNAs from a single gene

Table 1. Alternative nomenclature for the LIR gene family

Molecule	Ig repeats	ITIM-like motifs	References
LIR-1, ILT2, MIR7	4	4	Cosman et al. 1997; Samaridis and Colonna 1997; Wagtmann et al. 1997
LIR-2, ILT4, MIR10	4	3	Borges et al. 1997; Colonna et al. 1998; Wagtmann et al. 1997
LIR-3, ILT5, HL9	4	4	Borges et al. 1997; Colonna et al. 1998; Arm et al. 1997
LIR-4, ILT6, HM43	4	0	Borges et al. 1997; Colonna et al. 1997; Arm et al. 1997
LIR-5, ILT3, HM18	2	3	Borges et al. 1997; Cella et al. 1997; Arm et al. 1997
LIR-6a, LIR-6b	4,2	0,0	Borges et al. 1997
LIR-7, ILT1	4	0	Borges et al. 1997; Samaridis and Colonna 1997
LIR-8	4	2	Borges et al. 1997

quence similarities between LIRs and KIRs, we predict that LIRs devoid of ITIM sequences engage activatory signaling pathways. The same may be true for the murine PIR-A molecule.

3 LIR Distribution

The LIR cellular distribution is clearly distinct from that of KIRs. LIRs are broadly expressed in cells of both the myeloid and lymphoid lineage, while KIRs are restricted to NK and small subsets of T cells (LANIER and PHILLIPS 1996). LIR-1 is expressed in all $CD14^+$ monocytes, most of the $CD19^+$ B cells (97%), a low percentage of $CD3^+$ T cells (3%) and a variable percentage of $CD56^+$ NK cells (0%–30%) (our unpublished observations). High expression of LIR-1 is also observed in $CD33^+$, $CD14^-$, $CD16^-$, $HLA-DR^+$ peripheral blood dendritic cells. Colonna et al. (COLONNA et al. 1997) used a different antibody against ILT2/LIR-1 and describe the same general expression pattern. In addition, this group characterized the T cells expressing ILT2/LIR-1, as carrying either TCR-αβ or TCR-γδ chains.

LIR-2 has a more restricted cellular distribution than LIR-1. By flow cytometry analysis, LIR-2 is detected in all CD14$^+$ monocytes, and in a large percentage of CD33$^+$, CD14$^-$, CD16$^-$, HLA-DR$^+$ dendritic cells (our unpublished observations). LIR-2 expression is not observed in CD19$^+$ B cells, CD3$^+$ T cells, or CD56$^+$ NK cells. A similar pattern of expression for ILT3/LIR-5 was described by CELLA et al. (1997). ILT3/LIR-5 staining was observed in CD14$^+$ monocytes, and in CD14$^-$, HLA-DR$^+$ circulating dendritic cells, but no expression was detected in CD3$^+$ T cells, CD20$^+$ B cells, or CD16$^+$ NK cells.

The cellular distribution of the other LIR molecules has only been studied at the mRNA level. We used reverse transcription polymerase chain reaction and primers specific for each one of the LIR molecules to study the expression of these molecules in dendritic cells, monocytes, T cells, B cells, and NK cells (BORGES et al. 1997). LIR-3 and LIR-6a/b transcripts are found in monocytes and at lower levels in B cells. On the other hand, LIR-4 transcripts are most abundant in B cells and can be detected at lower levels in NK cells. A much more restricted expression pattern is observed with LIR-8, whose transcripts can only be detected in NK cells. LIR-7 transcripts were expressed at low to undetectable levels in all cells studied. More detailed distribution studies will be possible when additional LIR-specific antibodies become available.

4 LIR-1 and LIR-2 Bind to MHC Class I Molecules

Given the relationship between UL18 and MHC class I molecules, we examined the ability of fusion proteins between the LIRs' extracellular domains and human IgG1 Fc region, to bind to activated peripheral blood T cells, which express a high level of MHC class I. Only LIR-1 and LIR-2 bound to the activated T cells and this binding could be competed by the pan anti-MHC I antibody W6/32, implying that both these molecules were recognizing class I alleles (BORGES et al. 1997). Additionally, we isolated HLA-A2 and HLA-B44 cDNAs from an HSB-2 (a T lymphoblastoid leukemia cell line) cDNA expression library, using LIR-1-Fc binding as a screen (COSMAN et al. 1997).

To look in more detail at the binding specificity of LIR-1 and LIR-2, we used the B lymphoblastoid cell line 721.221, either untransfected or transfected with various MHC class I alleles. Both LIR-1 and LIR-2 showed a broad binding specificity to alleles of the HLA-A (A0101, A0301), HLA-B (B0702, B0801, B1501, B2702), and HLA-C (CW0304) haplotypes (FANGER et al. 1998). In addition to the classical class I antigens, ILT2/LIR-1 and ILT4/LIR-2 also bind to the non-classical class I antigen HLA-G1 (COLONNA et al. 1997). The range of HLA alleles recognized by the LIRs is much broader than that of the KIRs. Individual KIRs recognize HLA alleles of the A, B, or C allotypes. While it seems clear that most, if not all, of the KIRs, recognize class I molecules, evidence from our own and other laboratories suggests that most LIRs recognize ligands other than class I molecules. We have used the LIR-Fc fusion proteins to test binding to activated T cells and a large number of class I positive cell lines and consistently failed to detect any binding, with the exception of LIR-1 and LIR-2. In the few cases where we have observed binding, it could not be competed by an anti-MHC class I antibody, suggesting that antigens, other than class I, were being recognized. Possible binding partners for the LIRs might include rare MHC class I antigens, non-classical class I alleles, MHC class II molecules, and soluble Ig molecules. Of all the LIRs that we have cloned, only LIR-1 interacts with UL18 (BORGES et al. 1997), despite the high level of sequence identity among the LIR extracellular domains (for instance, LIR-2 and LIR-4 are 82% and 84% identical to LIR-1). This specificity suggests that LIR-1 has a non-redundant function that is targeted by hCMV.

5 LIR Signaling and Function

Five of the eight LIR molecules contain ITIM-like sequences in the cytoplasmic region (Fig. 1). ITIMs are present in inhibitory receptors such as KIRs and FcγRIIb, which block activation by other cell surface receptors in NK and B cells, respectively. Upon phosphorylation of the tyrosine residue, ITIMs recruit SH2 domain-containing phosphatases such as SHP-1 and SHIP, which then deliver an inhibitory signal (YOKOYAMA 1997). We have labeled the YxxL/V sequences in the LIR molecules as ITIM-like, since two of the four motifs do not conform to the consensus ITIM sequence I/VxYxxL/V (where x is any amino acid). For simplicity, we have numbered these ITIM-like sequences from 1 to 4, with motif 1 being the closest to the membrane. Motifs 2 and 3 have the sequence VxYxxL/V which fits the prototype ITIM sequence. On the other hand, motifs 1 and 4 do not have a hydrophobic residue at position –2 upstream of the tyrosine residue. Instead, this position is occupied by a polar amino acid, either an asparagine residue (motif 1) or a serine residue (motif 4). It is not known whether these motifs are involved in recruiting inhibitory signaling molecules. Since the sequences of motifs 1 and 4 are well conserved among the LIR molecules, it is likely that they might be involved in signaling mechanisms. Interestingly, the mouse PIR-B/p91 molecule has four ITIM-like sequences, but only one of them conforms to the consensus ITIM (HAYAMI et al. 1997; KUBAGAWA et al. 1997). The other three motifs lack a hydrophobic residue at position –2, which is either occupied by a glutamic acid or a serine residue. Evidence from yet another receptor system suggests that such motifs, or at least those preceded by a serine at position –2, could be involved in negative signaling. Mast cells express a C-type lectin receptor, MAFA (mast-cell function-associated antigen), which upon aggregation blocks Ig-E-induced activation of these cells (GUTHMANN et al. 1995a). The cytoplasmic domain of MAFA contains a single YxxL motif, which is preceded at position –2 by a serine residue (GUTHMANN et al. 1995b). It is not known yet whether this motif is involved in recruiting an inhibitory signaling molecule, but if so, that would support the notion that there is not an absolute requirement for a hydrophobic amino acid residue at position –2. The presence of different ITIM-like sequences in LIRs is probably required for the recruitment of different signaling molecules that might synergize to amplify the inhibitory signaling cascade. Since the same LIR molecule can be expressed in a wide variety of cells, it is also possible that the different motifs will be involved in recruiting different proteins expressed specifically in certain cell subsets.

In monocytes, LIR-1 and LIR-2 suppress signaling by the high affinity Fc receptor for IgG, CD64 (FANGER et al. 1998). Crosslinking CD64 on the surface of monocytes induces Ca^{2+} mobilization and a rapid increase in the levels of tyrosine phosphorylation of several proteins, including the FcR γ chain and the protein kinase Syk. However, when we used anti-LIR-1 or anti-LIR-2 $F_{(ab')}2$ antibody fragments to co-crosslink CD64 to either LIR-1 or LIR-2, we saw a marked inhibition of Ca^{2+} mobilization and overall tyrosine phosphorylation. Among the signaling proteins, we observed a significant decrease in the tyrosine phosphorylation level of both Syk and FcR γ chain. This inhibition appears to be LIR specific, since no detectable decreases in tyrosine phosphorylation could be seen when we crosslinked CD64 to either CD11c or CD14. Colonna and his collaborators (1997) have also demonstrated that the engagement of ILT2/LIR-1 on the surface of NK and T cells blocks cell-mediated cytotoxicity. In addition, this group has observed that co-engagement of ILT4/LIR-2 with either FcγRII or HLA-DR prevents Ca^{2+} mobilization in dendritic cells (COLONNA et al. 1998). Co-engagement of ILT3/LIR-5 with FcγRII or HLA-DR also blocks Ca^{2+} mobilization in antigen-presenting cells (CELLA et al. 1997). Interestingly, ILT3/LIR-5 is also involved in antigen capture and delivery to a processing compartment where class II loading occurs (CELLA et al. 1997). Signals for ILT-3 internalization might be encoded by the YxxV/L sequences present in the ITIM-like motifs, as these same sequences have been shown to be required for endocytosis of other proteins, including FcγRIIB2 (MIETTINEN et al. 1989). Mutation of the tyrosine residue into glycine blocks endocytosis mediated by FcγRIIB2 (DAERON et al. 1993). Since ITIM-like motifs are present in several LIR molecules, these molecules may also be involved in antigen capture by dendritic cells and monocytes. The antigen could then be deliver to a processing compartment for loading into MHC class I or II molecules.

The detailed inhibitory signaling pathways triggered by the LIRs are not yet known. Evidence from our own studies and those of Colonna's laboratory has implicated the tyrosine phosphatase SHP-1 as one of the enzymes likely to be involved in early signaling events triggered by the LIRs. In the B cell line CB23, SHP-1 associates with' tyrosine phosphorylated-LIR-1 after stimulation with sodium pervanadate (a general inhibitor of tyrosine phosphatases; COSMAN et al. 1997). In peripheral blood monocytes, SHP-1 co-immunoprecipitates with both LIR-1 and LIR-2, following sodium pervanadate treatment (FANGER et al. 1998). CELLA et al. (1997) have shown that ILT3/LIR-5 also recruits SHP-1 in monocytes, when the receptor is crosslinked by antibodies attached to the tissue culture plate. Since SHP-1 is required for signaling by other inhibitory receptors, including KIRs

(FRY et al. 1996) and the erythropoietin receptor (KLINGMULLER et al. 1995), it is likely that this enzyme is being recruited by the LIRs to deliver an inhibitory signal. Given the fact that only two of the LIR ITIM-like motifs contain a hydrophobic residue at position –2, as required for binding to the SH2 domains of SHP-1 (BURSHTYN et al. 1997), it is conceivable that the other LIR ITIM-like motifs might recruit other inhibitory signaling molecules.

The expression of LIR-1 and LIR-2 in monocytes and dendritic cells suggests that MHC class I molecules modulate a broad range of immune responses, which are not limited to NK and T cells as previously thought. The biological function for these inhibitory receptors for MHC class I antigens on monocytes and dendritic cells is not clear. They might raise the thresholds required for the activation of monocytes and dendritic cells. For instance, the presence of LIRs in monocytes might allow for the presence of low levels of self-reactive antibodies in circulation. These antibodies could be present constitutively or be induced during the generation of humoral responses against infectious agents that encode epitopes cross-reactive with self molecules. Tissues binding auto-antibodies might become targets for monocytes/macrophages through interactions between Fc receptors on the surface of the monocytes/macrophages and the auto-antibodies on the surface of cross-reactive cells. Triggering of the Fc receptors could potentially activate monocyte/macrophage cytotoxicity and up-regulate the secretion of pro-inflammatory cytokines such as TNF-α, IL-6, IL-8, IL-12, and IL-18. However, inhibitory signaling by LIR-1 and LIR-2 after engagement with MHC class I antigens on the tissues might prevent monocyte/macrophage activation.

In dendritic cells, the expression of LIR-1 and LIR-2 might be required to block, for instance, the expression of co-stimulatory molecules. This could be important to prevent activation of T cells that might react with self peptides expressed in the context of MHC class I molecules in dendritic cells. Since T cells are selected against self reactive epitopes in the thymus, a self peptide-MHC class I complex in peripheral tissues might only be able to bind with low affinity to T cell receptors expressed by a small subset of T cells. The presence of inhibitory receptors such as LIR-1 and LIR-2 in dendritic cells might prevent the activation of such self reactive T cells and generation of auto-immune reactions.

6 Possible Outcomes of the LIR-1 and UL18 Interaction

The significance of the LIR-1 interaction with UL18 is not known. Since LIR-1 is only expressed in a small subset of NK cells that varies from donor to donor (0%–30% of the total NK cells), it seems unlikely that CMV would target LIR-1 as an efficient way to block NK cytotoxicity. A recent report by REYBURN et al. (1997) suggests that UL18 expression on a CMV-infected cell engages CD94 on NK cells and prevents NK cytotoxicity. However, several lines of evidence contradict this hypothesis. We and others (LEONG et al. 1998) have been unable to detect any interactions between UL18 and CD94. When we used a UL18-Fc fusion protein to examine binding to NK cells, only small subsets of these cells were able to bind this protein, despite the wide expression of CD94 in these cells. In addition, all of the binding of UL18-Fc could be blocked by an antibody against LIR-1, suggesting that UL18 was recognizing only LIR-1 on the surface of NK cells (COSMAN et al. 1997). LEONG et al. (1998) have shown that several different cell lines that were either infected with CMV or transfected with UL18 are killed by NK cell clones expressing both KIRs and CD94. In this set of experiments, expression of UL18 consistently enhanced, rather than inhibited, killing and antibodies against CD94 did not affect NK cytotoxicity. Recent reports by several groups have clearly demonstrated that CD94/NKG2A interacts with HLA-E and does not have a broad class I binding specificity as previously thought (BRAUD et al. 1998; LEE et al. 1998). HLA-E molecules can only reach the cell surface when bound to leader peptides from certain class I molecules. In the Reyburn study (1997), 721.221 cells were transfected with a UL18 cDNA, but selected for the surface expression of β2-microglobulin and not UL18. As LEONG et al. (1998) suggested, it is possible that a variant of 721.221 cells expressing a higher level of endogenous HLA-E, and not UL18, was selected. In that case, HLA-E expression on the target cells could have engaged CD94 on the NK cells and blocked cytotoxicity.

Given the current evidence, we favor the hypothesis that CMV utilizes UL18 to modulate LIR-1 function in monocytes (Fig. 2). The interaction of UL18 and LIR-1 could either take place on a virus-infected cell interacting with a LIR-1-expressing cell or within the same cell. Expression of UL18 on a CMV-infected cell could engage LIR-1 on the surface of a monocyte and block signaling pathways responsible for triggering anti-viral functions. Blocking the secretion of cytokines such as IL-12, IL-15, and IL-18, could result in the failure of monocytes to activate NK cells. This in turn would limit the production of IFN-γ and compromise NK cytotoxicity towards CMV-infected cells. In mice, IFN-γ clearly limits CMV infection and blocks reactivation of the virus from latency (PRESTI et al. 1998). Since LIR-1 can block signaling by Fc receptors, another possible outcome of the LIR-1/UL18 interaction could happen later during the infection cycle. When antibodies against CMV proteins are generated by the host, they can potentially bind to viral proteins expressed on the surface of infected cells, trigger Fc receptors on monocytes and induce cytotoxicity. However, engagement of LIR-1 on the surface of monocytes could initiate an inhibitory signaling cascade that would block killing of the infected cell.

The LIR-1/UL18 interaction could also take place within the same cell. In healthy donors, peripheral blood monocytes are a viral reservoir for latent CMV (SODERBERG-NAUCLER et al. 1997). Since monocytes are one of the main cell types that express LIR-1, it is likely that both LIR-1 and UL18 could be expressed in CMV-infected monocytes. If that is the case, molecular interactions might take place and several outcomes could be hypothesized: 1) engagement of LIR-1 by UL18 could induce an inhibitory signaling pathway that could block anti-viral effector mechanisms in monocytes, possibly allowing the virus to establish latency, or 2) UL18 could sequester LIR-1 to facilitate full activation of a stimulatory

1. UL18 / LIR-1 interaction between two different cells.

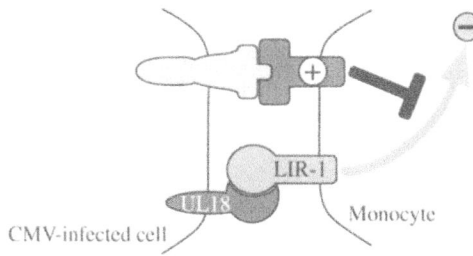

- Inhibition of cytokine release
- Inhibition of NK cell activation
- Inhibition of cytotoxicity

2. UL18 / LIR-1 interaction within the same cell.

A) Triggering of LIR-1 Signalling

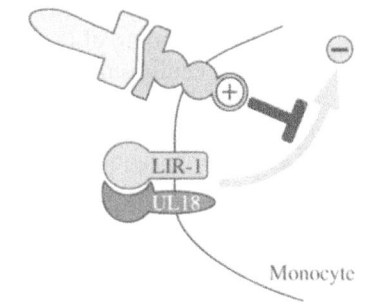

- Inhibition of monocyte activation
- Inhibition of cytokine production

B) Sequestration and inhibition of LIR-1 Signalling

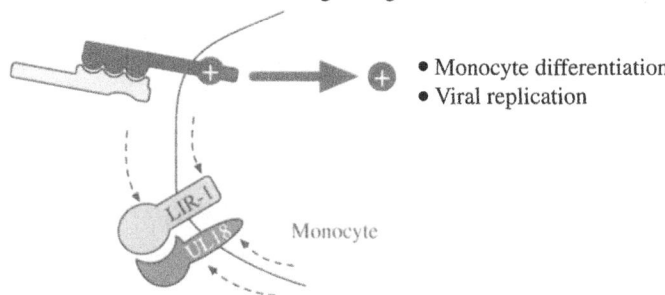

- Monocyte differentiation
- Viral replication

Fig. 2. Possible outcomes from UL18/leukocyte immunoglobulin-like receptor-1 (*LIR-1*) interactions. Interaction between LIR-1 and UL18 could take place either when a virus-infected cell contacts a LIR-1 expressing monocyte (*Panel 1*) or within a LIR-1 expressing monocyte that has been infected by CMV (*Panels 2 A,B*)

signaling pathway(s) that would be advantageous for viral replication or other viral function. For instance, studies have shown that, although CMV can infect myeloid progenitor cells (KONDO et al. 1994, 1996), the virus requires a more differentiated phenotype for efficient replication. In that case, reactivation of the virus from a latent stage in myeloid progenitor cells might require the activation of signaling pathways that induce cell differentiation. Sequestration or blockage of potentially inhibitory molecules such as LIR-1, might be required for efficient activation of these pathways.

7 Concluding Remarks

The understanding of the functional relationship between UL18 and LIR-1 could shed new light on deciphering the function of LIR-1 and other inhibitory receptors. The 1980s and 1990's have seen spectacular advances in the study of stimulatory receptors in the immune system and other tissues. Biologists are now realizing that most, if not all, of the cellular responses to external stimuli are regulated by the crosstalk between stimulatory and inhibitory pathways. We are at the beginning of a new century and likely a new era in the understanding of the function of inhibitory receptors and the intricate homeostatic mechanisms that ultimately dictate the outcome of cellular responses.

Acknowledgments. The authors gratefully acknowledge Anne Aumell for her valuable editorial assistance, and Gary Carlton for preparing the figures.

References

Ahn KS, Angulo A, Ghazal P, Peterson PA, Yang Y, Fruh K (1996) Human cytomegalovirus inhibits antigen presentation by a sequential multistep process. Proc Natl Acad Sci USA 93:10990–10995
Arm JP, Nwankwo C, Austen KF (1997) Molecular identification of a novel family of human Ig superfamily members that possess immunoreceptor tyrosine-based inhibition motifs and homology to the mouse gp49B1 inhibitory receptor. J Immunol 159:2342–9
Biron CA, Byron KS, Sullivan JL (1989) Severe herpesvirus infections in an adolescent without natural killer cells. N Engl J Med 320:1731–5
Borges L, Hsu ML, Fanger N, Cosman D (1997) A family of human lymphoid and myeloid Ig-like receptors, some of which bind to MHC class I molecules. J Immunol 159:5192–6
Braud VM, Allan DS, O'Callaghan CA, Soderstrom K, D'Andrea A, Ogg GS, Lazetic S, Young NT, Bell JI, Phillips JH, Lanier LL, McMichael AJ (1998) HLA-E binds to natural killer cell receptors CD94/NKG2 A, B and C. Nature 391:795–9
Browne H, Smith G, Beck S, Minson T (1990) A complex between the MHC class I homologue encoded by human cytomegalovirus and beta 2 microglobulin. Nature 347:770–2
Bukowski JF, Woda BA, Welsh RM (1984) Pathogenesis of murine cytomegalovirus infection in natural killer cell-depleted mice. J Virol 52:119–28
Burshtyn DN, Yang WT, Yi TL, Long EO (1997) A novel phosphotyrosine motif with a critical amino acid at position –2 for the SH2 domain-mediated activation of the tyrosine phosphatase SHP-1. J Biol Chem 272:13066–13072
Castells MC, Wu X, Arm JP, Austen KF, Katz HR (1994) Cloning of the gp49B gene of the immunoglobulin superfamily and demonstration that one of its two products is an early-expressed mast cell surface protein originally described as gp49. J Biol Chem 269:8393–8401

Cella M, Dohring C, Samaridis J, Dessing M, Brockhaus M, Lanzavecchia A, Colonna M (1997) A novel inhibitory receptor (ILT3) expressed in monocytes, macrophages, and dendritic cells involved in antigen processing. J Exp Med 185:1743–1751

Chapman TL, Bjorkman PJ (1998) Characterization of a murine cytomegalovirus class I major histo-compatibility complex (MHC) homolog: comparison to MHC molecules and to the human cyto-megalovirus MHC homolog. J Virol 72:460–6

Colonna M (1996) Specificity and function of immunoglobulin superfamily NK cell inhibitory and stimulatory receptors. Immunol Rev 155:127–133

Colonna M, Navarro F, Bellon T, Llano M, Garcia P, Samaridis J, Angman L, Cella M, Lopez-Botet M (1997) A common inhibitory receptor for major histocompatibility complex class I molecules on human lymphoid and myelomonocytic cells. J Exp Med 186:1809–18

Colonna M, Samaridis J, Cella M, Angman L, Allen RL, O'Callaghan CA, Dunbar R, Ogg GS, Cerundolo V, Rolink A (1998) Human myelomonocytic cells express an inhibitory receptor for classical and nonclassical MHC class I molecules. J Immunol 160:3096–100

Cosman D, Fanger N, Borges L, Kubin M, Chin W, Peterson L, Hsu ML (1997) A novel immuno-globulin superfamily receptor for cellular and viral MHC class I molecules. Immunity 7:273–82

Daeron M, Malbec O, Latour S, Bonnerot C, Segal DM, Fridman WH (1993) Distinct intracytoplasmic sequences are required for endocytosis and phagocytosis via murine Fc gamma RII in mast cells. Int Immunol 5:1393–401

de Wit TP, Morton HC, Capel PJ, van de Winkel JG (1995) Structure of the gene for the human myeloid IgA Fc receptor (CD89) J Immunol 155:1203–1209

Fahnestock ML, Johnson JL, Feldman RM, Neveu JM, Lane WS, Bjorkman PJ (1995) The MHC class I homolog encoded by human cytomegalovirus binds endogenous peptides. Immunity 3:583–90

Fanger N, Cosman D, Peterson L, Braddy S, Maliszewski C, Borges L (1998) The MHC Class I binding proteins, LIR-1 and LIR-2, inhibit Fc receptor-mediated signaling in monocytes. Eur J Immunol 28:3423–3434

Farrell HE, Vally H, Lynch DM, Fleming P, Shellam GR, Scalzo AA, Davis-Poynter NJ (1997) Inhi-bition of natural killer cells by a cytomegalovirus MHC class I homologue in vivo [see comments]. Nature 386:510–4

Fry AM, Lanier LL, Weiss A (1996) Phosphotyrosines in the killer cell inhibitory receptor motif of NKB1 are required for negative signaling and for association with protein tyrosine phosphatase 1 C. J Exp Med 184:295–300

Gilbert MJ, Riddell SR, Plachter B, Greenberg PD (1996) Cytomegalovirus selectively blocks antigen processing and presentation of its immediate-early gene product. Nature 383:720–2

Gold E, Nankervis GA (1976) Cytomegalovirus. in "Viral infections of humans: epidemiology and control" (A. S. Evans) (pp. 143–161) Plenum Press, New York

Guthmann, MD, Tal, M, Pecht, I (1995a) A new member of the C-type lectin family is a modulator of the mast cell secretory response. Int Arch Allergy Immunol 107:82–6

Guthmann MD, Tal M, Pecht I (1995b) A secretion inhibitory signal transduction molecule on mast cells is another C-type lectin. Proc Natl Acad Sci USA 92:9397–401

Hayami K, Fukuta D, Nishikawa Y, Yamashita Y, Inui M, Ohyama Y, Hikida M, Ohmori H, Takai T (1997) Molecular cloning of a novel murine cell-surface glycoprotein homologous to killer cell inhibitory receptors. J Biol Chem 272:7320–7327

Hengel H, Koopmann JO, Flohr T, Muranyi W, Goulmy E, Hammerling GJ, Koszinowski UH, Momburg F (1997) A viral ER-resident glycoprotein inactivates the MHC-encoded peptide trans-porter. Immunity 6:623–32

Klingmuller U, Lorenz U, Cantley LC, Neel BG, Lodish HF (1995) Specific recruitment of SH-PTP1 to the erythropoietin receptor causes inactivation of JAK2 and termination of proliferative signals. Cell 80:729–38

Kondo K, Kaneshima H, Mocarski ES (1994) Human cytomegalovirus latent infection of granulocyte-macrophage progenitors. Proc Natl Acad Sci USA 91:11879–83

Kondo K, Xu J, Mocarski ES (1996) Human cytomegalovirus latent gene expression in granulocyte-macrophage progenitors in culture and in seropositive individuals. Proc Natl Acad Sci USA 93:11137–42

Kubagawa H, Burrows P, Cooper M (1997) A novel pair of immunoglobulin-like receptors expressed by B cells and myeloid cells. Proc Natl Acad Sci USA 94:5261–5266

Lanier LL, Phillips JH (1996) Inhibitory MHC class I receptors on NK cells and T cells. Immunol. Today 17:86–91

Lee N, Llano M, Carretero M, Ishitani A, Navarro F, Lopez-Botet M, Geraghty DE (1998) HLA-E is a major ligand for the natural killer inhibitory receptor CD94/NKG2 A. Proc Natl Acad Sci USA 95:5199–204

Leong CC, Chapman TL, Bjorkman PJ, Formankova D, Mocarski ES, Phillips JH, Lanier LL (1998) Modulation of natural killer cell cytotoxicity in human cytomegalovirus infection: the role of endogenous class I major histocompatibility complex and a viral class I homolog. J Exp Med 187: 1681–7

Long E, Burshtyn D, Clark W, Peruzzi M, Rajagopalan S, Rojo S, Wagtmann N, Winter C (1996) Killer cell inhibitory receptors: diversity, specificity, and function. Immunol Rev 155:135–144

Maliszewski CR, March CJ, Schoenborn MA, Gimpel S, Shen L (1990) Expression cloning of a human Fc receptor for IgA. J Exp Med 172:1665–1672

Miettinen HM, Rose JK, Mellman I (1989) Fc receptor isoforms exhibit distinct abilities for coated pit localization as a result of cytoplasmic domain heterogeneity. Cell 58:317–27

Presti RM, Pollock JL, Dal Canto AJ, O'Guin AK, Virgin IH (1998) Interferon gamma Regulates Acute and Latent Murine Cytomegalovirus Infection and Chronic Disease of the Great Vessels. J Exp Med 188:577–88

Reyburn HT, Mandelboim O, Vales-Gomez M, Davis DM, Pazmany L, Strominger JL (1997) The class I MHC homologue of human cytomegalovirus inhibits attack by natural killer cells. Nature 386:514–7

Rojo S, Burshtyn DN, Long EO, Wagtmann N (1997) Type I transmembrane receptor with inhibitory function in mouse mast cells and NK cells. J Immunol 158:9–12

Samaridis J, Colonna M (1997) Cloning of novel immunoglobulin superfamily receptors expressed on human myeloid and lymphoid cells: Structural evidence for new stimulatory and inhibitory pathways. Eur J Immunol 27:660–665

Soderberg-Naucler C, Fish KN, Nelson JA (1997) Reactivation of latent human cytomegalovirus by allogeneic stimulation of blood cells from healthy donors. Cell 91:119–26

Suto Y, Maenaka K, Yabe T, Hirai M, Tokunaga K, Tadokoro K, Juji T (1996) Chromosomal localization of the human natural killer cell class I receptor family genes to 19q13.4 by fluorescence in situ hybridization. Genomics 35:270–272

Wagtmann N, Rojo S, Eichler E, Mohrenweiser H, Long EO (1997) A new human gene complex encoding the killer cell inhibitory receptors and related monocyte/macrophage receptors. Curr Biol 7:615–8

Wentworth BB, Alexander ER (1971) Seroepidemiology of infectious due to members of the herpesvirus group. Am J Epidemiol 94:496–507

Wiertz EJ, Tortorella D, Bogyo M, Yu J, Mothes W, Jones TR, Rapoport TA, Ploegh HL (1996a) Sec61-mediated transfer of a membrane protein from the endoplasmic reticulum to the proteasome for destruction [see comments]. Nature 384:432–8

Wiertz EJHJ, Jones TR, Sun L, Bogyo M, Geuze HJ, Ploegh HL (1996b) The human cytomegalovirus US11 gene product dislocates MHC class I heavy chains from the endoplasmic reticulum to the cytosol. Cell 84:769–779

Yokoyama WM (1997) What goes up must come down: the emerging spectrum of inhibitory receptors. J Exp Med 186:1803–8

Zhang G, Young JR, Tregaskes CA, Sopp P, Howard CJ (1995) Identification of a novel class of mammalian Fc gamma receptor. J Immunol 155:1534–1541

Paired Immunoglobulin-like Receptors
of Activating and Inhibitory Types

H. Kubagawa[1], M.D. Cooper[2], C.C. Chen[3], L.H. Ho[2], T.L. Alley[4], V. Hurez[4],
T. Tun[2], T. Uehara[5], T. Shimada[4], and P.D. Burrows[4]

1 Introduction

A distinguishing feature of the immune system is the ability to respond rapidly and efficiently to pathogenic insult. Cellular immune responses, which may include proliferation, elaboration of cytokines, and cytotoxicity, are initiated by receptor–ligand interactions. To maintain immune system homeostasis, the responses initiated by activating receptors are normally counterbalanced by signals propagated through ligation of corresponding inhibitory receptors. Protein motifs that nucleate the activating or inhibitory cascade following receptor ligation have been identified

[1] Department of Pathology, Division of Developmental and Clinical Immunology, University of Alabama, Birmingham, AL 35294-3300, USA
[2] Howard Hughes Medical Institute, Birmingham, AL 35294-300, USA
[3] Department of Microbiology, Division of Developmental and Clinical Immunology, University of Alabama, Birmingham, AL 35294-3300, USA
[4] Department of Medicine, Division of Developmental and Clinical Immunology, University of Alabama, Birmingham, AL 35294-3300, USA
[5] Department of Medicine, Division of Developmental and Clinical Immunology, University of Alabama, Birmingham, AL 35294-3300, USA

in the cytoplasmic tails of many immune system molecules (WEISS and SCHLES-SINGER 1998). The immunoreceptor tyrosine-based activation motif (ITAM), with the consensus sequence of $D/E-x_7-D/E-x-x-Y-x-x-L/I-x_7-Y-x-x-L/I$, was first identified in the signal transducing components of the antigen receptor complexes on B and T lymphocytes (RETH 1989; CAMBIER 1995). The immunoreceptor tyrosine-based inhibitory motif (ITIM), with the consensus sequence of $I/V/L/S-x-Y-x-x-V/L$, which is the focus of this volume, was originally described in the cytoplasmic tails of the low affinity Fc receptor for IgG antibodies (FcγRIIB) on B cells and the killer inhibitory receptors (KIR) on NK cells (RAVETCH 1994; DAÉRON 1997; VÉLY and VIVIER 1997).

We have recently identified a family of transmembrane proteins, designated as paired immunoglobulin-like receptors (PIR), that have the potential to transmit activation or inhibitory signals to hemopoietic cells through the ITAM or ITIM, respectively. In this article we review the genetic and biochemical features, cellular distribution, and functional aspects of the *Pir* gene family.

2 Identification of the *Pir* Gene Family

We and others have characterized the structure and function of an Fc receptor for IgA (FcαR) that is expressed by human myeloid lineage cells, namely neutrophils, monocyte/macrophages and activated eosinophils. Based upon sequence homology, genomic organization and chromosomal localization, the gene encoding the human FcαR is the most distant relative within the FcR gene family (see review by MORTON et al. 1996). When the human FcαR cDNA probe was used in mouse DNA blot analysis in an attempt to isolate the murine FcαR gene, it cross-hybridized to two EcoRI-digested DNA fragments of ~3.6kb and ~3.4kb under relatively high stringency conditions. These DNA fragments were then cloned from a BALB/c mouse genomic library, and nucleotide sequence analysis revealed that they were closely related and appeared to contain several exons with limited homology to the human FcαR. As these genes were selectively expressed in bone marrow and spleen, exon-containing DNA fragments from genomic clones were used as probes to screen a BALB/c splenic cDNA library. This led to the identification of PIR gene family, whose members are related to the gene family encoding activating and inhibitory receptors expressed by human NK cells (KUBAGAWA et al. 1997). Employing a similar cloning strategy, Takai and his colleagues independently identified a PIR-B cDNA as p91 (HAYAMI et al. 1997). The PIR family members share sequence similarity (40%–60%) with other genes encoding the human FcαR, the mouse gp49 on mast cells, the bovine FcγR on alveolar macrophages, and the recently identified human Ig-like transcripts (ILT)/leukocyte Ig-like receptors (LIR)/monocyte–macrophage Ig-like receptors (MIR) on various hemopoietic cell types (see other chapters in this volume).

3 PIR-A and PIR-B

Nucleotide sequences of multiple PIR cDNA clones predict the existence of two types of PIR molecules which we have termed PIR-A (A for activating) and PIR-B (B for braking or inhibitory). Both are type I transmembrane proteins with very similar (>92% homology) ectodomains, each containing six Ig-like domains, the second and fourth of which are V sets, and the others are C2 sets. PIR-A and PIR-B have distinctive membrane proximal, transmembrane, and cytoplasmic regions (Fig. 1). The predicted PIR-A protein, with an estimated core size of 73kDa, has a short cytoplasmic tail without recognizable functional motifs, but contains a charged arginine residue in its transmembrane region. This suggests potential non-covalent association of PIR-A with an additional transmembrane protein (see below). In contrast, the PIR-B protein, with an estimated core size of 91kDa, has a typical uncharged transmembrane region and a long cytoplasmic tail with multiple candidate ITIMs.

While multiple independent PIR-B cDNAs had the identical nucleotide sequence, PIR-A cDNA clones exhibited sequence variations in their ectodomains. Some of the nucleotide changes were silent but others resulted in amino acid differences. These were distributed throughout the extracellular (EC) region, but were predominantly seen in the first four EC domains. Within the V-type EC2 and EC4 domains, such amino acid changes were often clustered in regions analogous to the

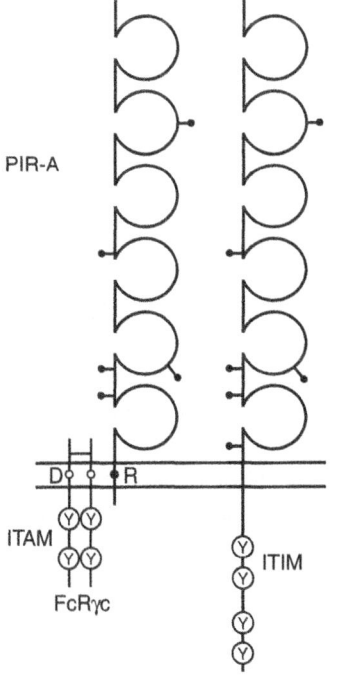

PIR-A

PIR-B

Fig. 1. Predicted protein structure of PIR-A and PIR-B. Both PIR-A (*left*) and PIR-B (*right*) cDNA clones encode type I transmembrane proteins with similar ectodomains, but distinctive transmembrane and cytoplasmic regions. The ectodomain has five or six potential sites for N-linked glycosylation. The predicted PIR-A has a short cytoplasmic tail and a positively charged arginine (*R*) residue in the transmembrane region, which is non-covalently associated with a negatively charged aspartic acid (*D*) in the transmembrane domain of the disulfide-linked homodimer of the Fc receptor common γ chain (*FcRγc*) carrying immunoreceptor tyrosine-based activation motifs (*ITAM*). In contrast, the PIR-B protein has a typical uncharged transmembrane region and a long cytoplasmic tail with the immunoreceptor tyrosine-based inhibitory motifs (*ITIM*)

CDR1 and CDR2 loops of Ig variable regions. In an ongoing analysis we have so far identified seven different PIR-A cDNAs (A1–A7) and one PIR-B cDNA. Results from DNA blot analysis using PIR-A- and PIR-B-specific probes have suggested the presence of multiple *Pira* genes but only one *Pirb* gene in the mouse genome. In addition to subtle sequence variations, the PIR-A cDNAs displayed size heterogeneity due either to the alternative usage of four possible polyadenylation signals or to the alternate splicing of EC exons. Unlike PIR-A, neither alternate polyadenylation signals nor splice variants were observed among five PIR-B cDNAs.

Restricted tissue and cellular expression of the *Pir* genes was evident from RNA blot analysis. Three major PIR transcripts of ~3.5, ~2.7 and ~2.5kb were detected with a common EC probe in bone marrow and spleen, but not in the thymus, brain, kidneys, intestine, skin, heart, and skeletal muscles. Weak signals were also observed for adult liver and lungs, possibly reflecting transcripts either in cells derived from the circulation or in Kupffer cells and alveolar macrophages. Transcripts of the same sizes were expressed by several B and myeloid cell lines, but not by pro-B, pre-B, plasmacytoma, T, and fibroblast cell lines in levels detectable by RNA blot analysis. The coordinate or "paired" expression of the *Pira* and *Pirb* genes was demonstrated by reverse transcriptase-dependent polymerase chain reaction (PCR) analysis of clonal B and myeloid cell lines using primers specific for each receptor type. The PCR products of PIR-B and PIR-A were selectively identified in pro-B, B, myeloid and mast cell lines as well as in normal B220$^+$ B cells, Mac-1$^+$ macrophages, and Gr-1$^+$ granulocytes.

4 Genomic Organization of the *Pirb* Gene

The *Pir* gene family was mapped to the proximal end of chromosome 7 where the intracisternal A-particle proviral elements (Iapls3–4), endogenous polytropic murine leukemia provirus (Pmv-4), avian reticuloendotheliosis viral oncogene-related B (Relb), and dehydroepiandrosterone sulfotransferase (Std) genes are clustered (KUBAGAWA et al. 1997; YAMASHITA et al. 1998). Interestingly, this region is syntenic with the human chromosome 19q13 region where the FcαR, KIR, and ILT/LIR/MIR genes are mapped (WAGTMANN et al. 1997).

Information about the genomic organization of the entire *Pir* gene family will be essential for understanding the genetic diversity of *Pira* and the regulation of *Pira* and *Pirb* gene expression. However, we have first focused on the genomic organization of the single copy *Pirb* gene. Nucleotide sequence analysis of three overlapping clones from a 129SV mouse genomic library has revealed that the *Pirb* gene consists of 15 exons and spans ~8kb (ALLEY et al. 1998; YAMASHITA et al. 1998). The first exon (5'UT/S1) encodes the 5' untranslated (UT) region, the ATG translation start site and approximately half of the signal peptide. Exon 2 (S2) is 36bp long and encodes the remainder of the signal peptide. The finding of two separate exons for the signal peptide of the *Pirb* gene is a common feature among genes encoding the ligand-binding chain of FcεRI, FcαR and all members of FcγRs (including human, mouse, rat, and bovine), the human KIR, and the mouse gp49. Exons 3 to 8 (EC1 to EC6) each encode a single Ig-like EC domain. Exon 9 (ECmp) is 52bp long and encodes the membrane proximal EC region. Exon 10 (TM) encodes the remaining EC region, the entire transmembrane segment, and a part of the cytoplasmic region. Exons 11 to 15 (CY1 to CY5/3′UT) encode the remaining cytoplasmic and 3′ UT regions. All exon/intron boundaries conform to the GT-AG rule and are phase I (i.e., they occur after the first nucleotide of a triplet codon), except for the boundaries of exons encoding cytoplasmic regions. DNA sequences of the exons in 129SV mice exhibited ~98% identity to the corresponding sequences in the PIR-B cDNA derived from a BALB/c mouse, indicating allelic differences between these two inbred strains. A microsatellite composed of the trinucleotide repeat AAG was identified in the intron between exons 9 and 10 and might provide a useful marker for studying population genetics. Inspection of the immediate 5′ flanking region (~1.8kb) failed to identify TATA- or CAAT-like promoter sequences, but revealed multiple initiator (Inr) sites – functional analogues of the TATA-promoter (ERNST and SMALE 1995) – at 28, 41, and 146bp upstream of the ATG translational start site of *Pirb*. A number of DNA sequence motifs that may bind lymphocyte-specific or ubiquitous transcriptional factors (FAISST and MEYER 1992), including E2 A, TCF-1, PU.1, Ets-1, NF-IL6, PEA3, AP1, and AP2, were also identified within the 5′ flanking region of *Pirb* gene.

5 Biochemical Nature of PIR Proteins

A histidine-tagged recombinant protein consisting of the two amino-terminal EC1/EC2 Ig-like domains of PIR-A1 was produced in *Escherichia coli*. This region was selected to immunize rats and rabbits because of its sequence identity with PIR-B. The resultant rat monoclonal antibody (mAb) of a γ1κ isotype was found to recognize a common epitope present on native PIR-A and PIR-B, whereas rabbit polyclonal antibodies reacted additionally with shared hidden epitopes on both molecules. These non-discriminating antibodies were used in immunoprecipitation analysis to define the molecular nature of PIR-A and PIR-B on stable PIR transfectants, representative hematopoietic cell lines and splenocytes (KUBAGAWA et al. 1999). While the PIR-B transfected fibroblasts expressed a transgene product of ~120kDa glycoprotein on their cell surface, the PIR-A1 transfected fibroblasts produced a glycoprotein of ~85 kDa, but failed to express the PIR-A molecules on the cell surface (see below). Both molecular species were expressed on the cell surface of clonal B and myeloid cell lines, and on splenocytes (Fig. 2), consistent with the coordinate or "paired" expression of PIR-A and PIR-B transcripts. Slightly higher molecular masses (~5kDa) were evident under reducing conditions, in agreement with predicted intradisulfide linkages of the six Ig-like EC domains. N-glycanase treatment of these cell surface PIR molecules reduced their apparent

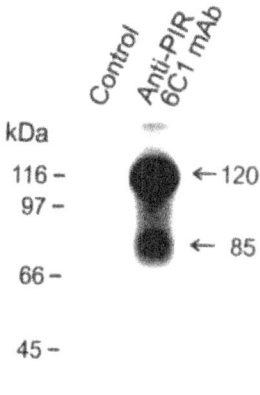

kDa

116 –   ←120
97 –

            ← 85

66 –

45 –

31–

Fig. 2. Immunoprecipitation analysis of cell surface PIR molecules. Cell surface proteins on splenocytes were labeled with ^{125}I by lactoperoxidase, solubilized in 1% NP-40, and added into wells precoated with anti-PIR or isotype-matched control monoclonal antibodies. The bound materials were resolved on SDS-10% PAGE under reducing conditions. The same results were obtained using B (X16C8.5) and macrophage (WEHI3) cell lines

molecular masses by 10–15kDa, in keeping with the presence of five or six potential N-linked glycosylation sites in the PIR-A and PIR-B receptors. The cell surface PIR-A and PIR-B molecules were thus defined as glycoproteins of ~85kDa and ~120kDa, respectively. The ~120kDa estimate for PIR-B agrees with that obtained using a rabbit antiserum against a p91/PIR-B cytoplasmic peptide (HAYAMI et al. 1997).

6 Association of the Fc Receptor Common γ Chain with PIR-A Receptors

When mouse LTK fibroblasts were transfected with the PIR-A1 or PIR-B expression vector, we observed an interesting difference in the resulting expression pattern. While the anti-PIR mAb recognized the cell surface receptors on PIR-B transfectants, it failed to identify cell surface molecules on PIR-A1 transfectants, even though PIR-A1 proteins were produced intracellularly. This difference implied a requirement for additional membrane-bound protein(s) for PIR-A cell surface expression, a possibility also suggested by the presence of a charged arginine residue in the transmembrane region of the predicted PIR-A1 protein. In theory, membrane proteins in which the transmembrane segment contains a negatively charged residue (aspartic acid or glutamic acid) are potential candidates for association with the PIR-A1 chain. These include FcR common γ chain (FcRγc), CD3ζ, 4-transmembrane spanning superfamily members (e.g., CD9, CD37, CD53, CD63), 7-transmembrane spanning or rhodopsin superfamily members (e.g., C5aR), and a recently identified dendritic cell-associated protein (DAP-12; LANIER et al. 1998). Among these, FcRγc, CD3ζ and DAP-12 are unique in that they contain ITAMs in their cytoplasmic regions.

Evidence for association of FcRγc with PIR-A came from the following observations. Firstly, the disulfide linked FcRγc homodimer (∼16kDa) was coprecipitated with the PIR molecules from digitonin lysates of splenocytes as determined by immunoblotting analysis with anti-FcRγc antibodies. This noncovalent association, like that of the FcγRI/FcRγc and FcαR/FcRγc complexes, did not readily dissociate even in NP-40, suggesting a relatively stable interaction. Secondly, the stable transfection of the FcRγc expression vector into the PIR-A1-producing fibroblasts induced the cell surface expression of PIR-A1 receptors. Thirdly, in collaboration with Jeffrey V. Ravetch (Rockefeller University, NY), splenocytes from wild-type mice were found to express both PIR-A and PIR-B on the cell surface, whereas only PIR-B was identified on the cell surface of splenocytes from FcRγc-deficient mice. In both groups of mice, however, PIR-A and PIR-B molecules were easily detected intracellularly. These findings thus indicate that, unlike PIR-B, PIR-A requires ITAM-bearing FcRγc chains for cell surface expression; and they imply that clustering of PIR-A receptor complexes by ligand ligation may transmit an activation signal to cells via the ITAMs of the FcRγc.

7 Cellular Distribution of PIR Proteins

When bone marrow cells from adult BALB/c mice were incubated with the fluorochrome-labeled anti-PIR mAb, ∼80% of the nucleated cells were stained. Most of the PIR$^+$ cells were myeloid lineage cells, but some were lymphocytes. By multicolor immunofluorescence assays, bone marrow cells bearing the Gr-1 and Mac-1 (CD11b, CR3) myelomonocytoid antigens were found to express the PIR antigen at variable levels that correlated with Gr-1 expression levels. Examination of the morphological features of isolated PIR^{lo}/Gr-1^{lo} and PIR^{hi}/Gr-1^{hi} fractions indicated an increase in PIR and Gr-1 expression with progression of granulocyte differentiation. CD19$^+$ B-lineage cells in bone marrow expressed PIR at lower levels relative to the myeloid cells, and most of the PIR$^+$ B-lineage cells were IgM$^+$ B cells. Pro-B cells were negative for PIR, while pre-B cells expressed PIR on their cell surface at relatively low levels, thus indicating a gradual increase in PIR expression as a function of B-lineage differentiation. Bone marrow-derived mast cells also expressed PIR on their cell surface. PIR was not detected on erythroid lineage cells (Ter119$^+$) in the bone marrow or on CD3$^+$ thymocytes from newborn or adult BALB/c mice. In contrast, thymic dendritic cells (MHC class II$^+$, CD11c$^+$, CD8$^+$, CD3$^-$, CD4$^-$) were found to express PIR at variable levels.

Splenic B cells (CD19$^+$) and macrophages (Mac-1$^+$) also expressed the PIR antigen on their cell surface, whereas NK cells (DX5$^+$) and T cells (CD3$^+$) did not (Fig. 3), consistent with the expression patterns noted for PIR-A and PIR-B transcripts in these cell types. Splenic dendritic cells (MHC class II$^+$, CD11c$^+$, CD8$^{+/-}$, CD19$^-$, Mac-1$^-$) also expressed PIR at low to high levels. PIR expression was higher on macrophages than on splenic B cells, the latter exhibiting variable

PIR levels. Evaluation of splenic B cell subpopulations indicated that the marginal zone B cells (B220$^+$, CD21^{high}, CD23$^{low/-}$) expressed higher PIR levels than newly formed (B220$^+$, CD21$^-$, CD23$^-$) and follicular B cells (B220$^+$, CD21^{med}, CD23$^+$). Moreover, the B1 cell populations (CD19$^+$/CD5$^+$ or CD19$^+$/CD5$^-$/Mac-1$^+$) in peritoneal lavage expressed higher levels of PIR than did the B2 cells (CD19$^+$, CD5$^-$, Mac-1$^-$). Consistent with the suggestion that B cell activation may enhance PIR expression, cell surface levels of PIR were upregulated following LPS stimulation of splenic B cells. Macrophage PIR expression was similarly enhanced by LPS stimulation.

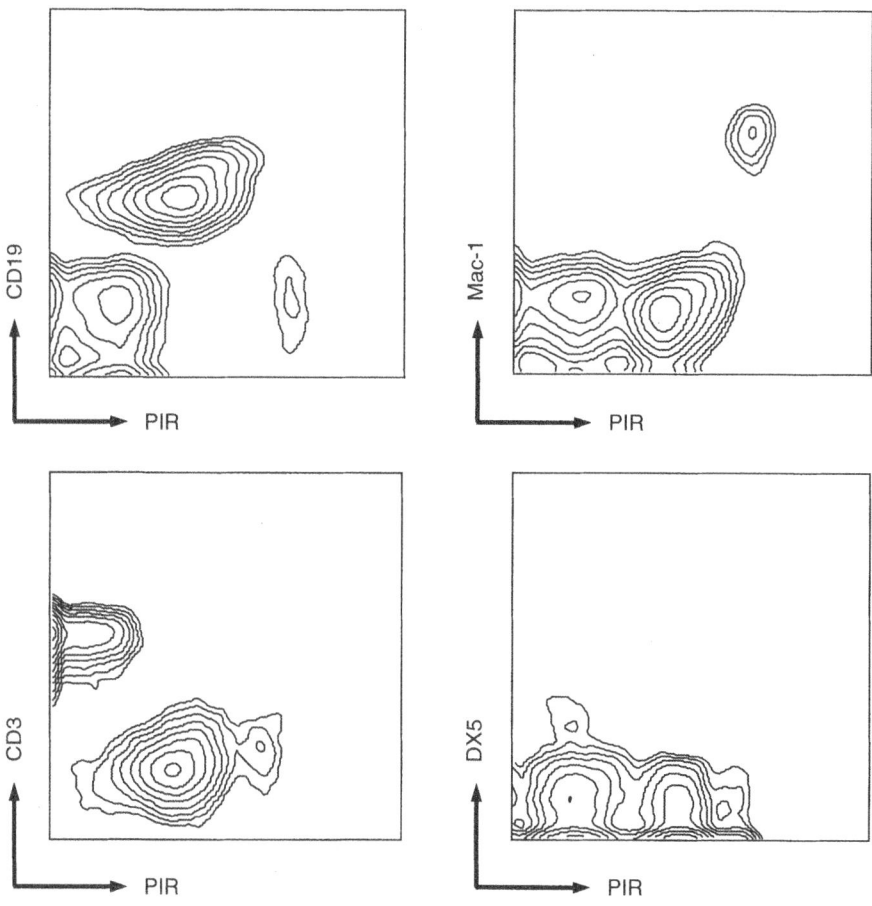

Fig. 3. Surface paired immunoglobulin-like receptors (PIR) expression by splenic B cells and macrophages. Splenocytes of adult BALB/c mice were incubated first with aggregated human IgG to block Fcγ receptors (FcγR), then with phycoerythrin-labeled anti-PIR and fluorescein isothiocyanate-labeled mAbs specific for CD19, Mac-1, CD3, or DX5. The stained cells with light scatter characteristics of lymphoid and large mononuclear cells were analyzed by flow cytometry. Note that CD19$^+$ B cells express PIR at variable intensity, while Mac-1$^+$ macrophages express PIR at relatively high intensity

8 Inhibitory Function of PIR-B Receptor

Since PIR-A and PIR-B are simultaneously expressed by B cells and macrophages, we anticipated that evaluation of the functional consequences of their cross-linkage using non-discriminative anti-PIR mAb [F(ab')$_2$ fragments] would be uninformative. Surprisingly, this was not the case. While ligation of PIR-A/PIR-B alone did not affect intracellular Ca^{2+} levels of splenic B cells, coligation of B cell receptor (BCR) complexes with PIR-A/PIR-B inhibited the BCR ligation-induced Ca^{2+} mobilization (BLÉRY et al. 1998). This inhibition was observed in the presence of EGTA and in Ca^{2+} free media, indicating that the PIR-B acts at least on Ca^{2+} release from intracellular pools. The PIR-B receptor may thus exert a dominant inhibitory effect on the BCR-mediated Ca^{2+} response of activated B cells.

To dissect the structural basis for the PIR-B inhibitory function, we have employed several different approaches in collaboration with Eric Vivier, Mathieu Bléry, and Frédéric Vély (Marseille, France). The predicted PIR-B protein has a long cytoplasmic tail (178 amino acids) containing five tyrosine residues, four of which – Y-713 (SLYASV), Y-742 (ETYAQV), Y-794 (VTYAQL), and Y-824 (SVYATL) – conform to the ITIM-like sequence motif. The fifth tyrosine, Y-770 (NTEYEQ), does not match the ITIM consensus sequence (Y-numbers indicate the position from the initiation methionine of PIR-B). To determine if these ITIM-like motifs have the potential to bind to the protein tyrosine phosphatases SHP-1 and SHP-2, or to the polyphosphate inositol 5-phosphatase SHIP, we generated synthetic phosphopeptides (15 mer) corresponding to the above tyrosine-containing amino acid stretches. When the phosphopeptide-coated agarose beads were incubated with lysates of an ^{35}S-Met/^{35}S-Cys labeled B cell line, two prominent bands, the 64kDa SHP-1 and the 68kDa SHP-2, were selectively precipitated by the phosphopeptides corresponding to the two carboxy-terminal tyrosine residues (Y-794 and Y-824; BLÉRY et al. 1998). Two faint bands of 68kDa (SHP-2) and 85kDa were also precipitated by the Y-713 phosphopeptide, and the molecular identification of the latter protein remains unresolved. None of the phosphopeptides bound SHIP ($\sim$145kDa).

Surface plasmon resonance assays were also conducted to measure the binding of phosphopeptide-coated microchips to the recombinant soluble SH2 domains of SHP-1, SHP-2 and SHIP (BLÉRY et al. 1998). Both Y-794 and Y-824 phosphopeptides bound to the SH2 domain of SHP-1 and SHP-2. The latter phosphopeptide also bound weakly the SH2 domain of SHIP. The Y-713 phosphopeptide exhibited only minimal binding with SHP-2, but none of the other phosphopeptides showed significant binding to the phosphatase SH2 domains.

To test the in vivo relevance of these in vitro data, we made constructs for a chimeric molecule (FcγRIIB/PIR-B) composed of the extracellular and transmembrane regions of murine FcγRIIB and the entire cytoplasmic domain of PIR-B. The wild-type or mutant (Y794F, Y824F and Y794F/Y824F) constructs were transfected into the rat basophilic leukemia cell line (RBL-2H3) to obtain stable transformants. Coaggregation of wild-type chimeric receptor with the endogenous FcεRI complex reduced IgE-mediated serotonin release, consistent with the hypothesis that PIR-B serves as an inhibitory receptor (BLÉRY et al. 1998). No inhibition was observed when the chimeric receptor and the FcεRI were independently aggregated, suggesting the importance of close proximity of both receptors for inhibition. The PIR-B cytoplasmic domain was tyrosine phosphorylated upon coengagement of chimeric receptor and FcεRI, as well as after ligation of chimeric receptor alone. SHP-1 was recruited to the PIR-B cytoplasmic domain upon coaggregation of both receptors, although weak association of SHP-1 with PIR-B was also observed when the chimeric receptor alone was aggregated. No association of the PIR-B cytoplasmic domain with SHP-2 or SHIP could be demonstrated in these studies. Mutation analysis revealed that only the double mutant (Y794F/Y824F) chimeric receptor efficiently abrogated the PIR-B inhibitory effect, suggesting redundancy of the PIR-B ITIMs. It should be noted that even this double mutant molecule was still capable of partial inhibition of FcεRI-induced activation, suggesting a weak inhibitory role of Y-713.

Kurosaki and his colleagues have made similar chimeric receptor (FcγRIIB/PIR-B) constructs and transfected them into the FcγRIIB-negative mouse B cell line (IIA1.6) or the chicken B cell line (DT40) deficient for SHP-1, SHP-2, or both (MAEDA et al. 1998). They found an inhibitory function for the PIR-B cytoplasmic domain in BCR ligation-mediated Ca^{2+} mobilization, especially Ca^{2+} release from intracellular stores, when the chimeric receptor and the BCR were coligated. The carboxy-terminal tyrosine residues Y-794 and Y-824, particularly the former, were found to contribute to the inhibition, results being consistent with ours. The DT40 transfection study, however, led to the different conclusion that both SHP-1 and SHP-2 phosphatases were required for the PIR-B-mediated inhibitory signal. The basis for this discrepancy (SHP-1 only versus SHP-1 and SHP-2) remains unclear, but may be related to the use of different cell types (rat RBL basophils versus chicken DT40 B cells) and different assays (FcεRI-mediated serotonin release versus BCR-mediated Ca^{2+} mobilization). In addition to B cells and mast cells, in macrophages PIR-B was found to associate with SHP-1 (TIMMS et al. 1998).

9 Human Counterparts

The evolutionary conservation of the *Pir* gene family was indicated by a zoo blot analysis employing DNA of human, monkey, cow, dog, mouse, rat, rabbit, chicken, and yeast origin with a common EC probe, where all samples, except for the yeast DNA, cross-hybridized with the mouse probe. When the same mouse probe was used in Southern blot analysis of human DNA digested with common restriction enzymes, multiple discrete fragments (at least 11) were found to cross-hybridize, suggesting that the human PIR counterparts also comprise a multigene family. Some of these cross-hybridizing fragments were cloned from a human genomic library, and the corresponding cDNA clones were isolated from a blood mononuclear cell library. Nucleotide sequence analysis revealed that the human PIR homologues were very similar or identical to members of the ILT/LIR/MIR gene family. This gene family was recently identified by at least four other groups of investigators, each using a unique cloning strategy (SAMARIDIS and COLONNA 1997; COSMAN et al. 1997; WAGTMANN et al. 1997; ARM et al. 1997; see their chapters in this volume).

Intriguing similarities and differences emerge from the comparison of mouse PIR and human ILT/LIR/MIR (Table 1). Like PIR, ILT/LIR/MIR is composed of two types, inhibitory and activating, based upon the existence of ITIM-like motifs in the long cytoplasmic tail of one isoform and a charged residue in the transmembrane domain of the other isoform with a short cytoplasmic tail. Unlike PIR, ILT/LIR/ MIR molecules have, at most, four extracellular Ig-like domains. In addition, sequence analysis reveals multiple receptors of the inhibitory type and a few receptors of the activating type for the ILT/LIR/MIR family. Sequence similarity between PIR and ILT/LIR/MIR is domain-dependent in that PIR EC 3, 2, 5 and 6 domains share the greatest homology (44%–64%) with ILT2 EC 1, 2, 3, and 4 domains, respectively. This human gene family has sequence similarity to human FcαR and KIR, mouse gp49, and bovine FcγR. ILT/LIR/MIR is expressed not only by B cells, myeloid cells, and dendritic cells, but also by NK cells and a minor subpopulation of T cells; and the inhibitory- and activating-type receptors are not always coordinately expressed. Certain members (LIR1/ILT2) of this receptor family are known to bind either classical or nonclassical MHC class I antigens (COSMAN et al. 1997; BORGES et al. 1997; COLONNA et al. 1998). Whether the ILT/LIR/MIR is indeed the human PIR counterpart remains to be elucidated. If so, we may learn important lessons from species differences with regard to the function of PIR molecules.

10 Concluding Remarks

A new family of paired immunoglobulin-like receptors, PIR-A (A for activating) and PIR-B (B for braking or inhibitory), has been identified in mice. The PIR

Table 1. Comparison of mouse PIR and human ILT/LIR/MIR

Feature	Mouse PIR	Human ILT/LIR/MIR
Multigene family	Yes	Yes
Activating	Multiple	Few
Inhibitory	Single	Multiple
Multiple RNA transcripts	Yes	Yes
Immunoglobulin domains	6	4
Paired expression	Yes	Not always
Cellular distribution	B, myeloid, DC	B, myeloid, DC, NK, T
Ligand	Unknown	Unknown/MHC class I

PIR, paired immunoglobulin-like receptors; ILT, Ig-like transcripts LIR, leukocyte Ig-like receptors; MIR, monocyte–macrophage Ig-like receptors family members share sequence similarity with human FcαR and KIR, mouse gp49, bovine FcγR, and the recently identified human ILT/LIR/MIR. The multiple *Pira* genes and a single *Pirb* gene are located at the proximal end of mouse chromosome 7, syntenic with the human chromosome 19q13 region where the FcαR, KIR and ILT/LIR/MIR genes are mapped; and the exon/intron organization of these genes is similar to that of the *Pirb* gene. PIR-A and PIR-B are type I transmembrane proteins with very similar ectodomains (> 92% homology), each containing six Ig-like domains, the second and fourth of which are V sets; and the others are C2 sets. PIR-A and PIR-B, however, have distinctive membrane proximal, transmembrane, and cytoplasmic regions. The predicted PIR-A protein has a short cytoplasmic tail and a charged arginine residue in its transmembrane region, whereas the PIR-B protein has a typical uncharged transmembrane region and a long cytoplasmic tail with multiple ITIMs. Monoclonal and polyclonal antibodies produced against a recombinant PIR protein identify the cell surface glycoproteins of ~85kDa and ~120kDa as PIR-A and PIR-B, respectively, on splenocytes and representative B- and myeloid-cell lines. The ITAM-bearing FcRγc is required for the cell surface expression of PIR-A molecules, and thus ligand-induced clustering of PIR-A receptor complexes may transmit an activation signal to cells via the ITAMs of the FcRγc. In contrast, the inhibitory function of the PIR-B cytoplasmic tail is evident for BCR-mediated Ca^{2+} mobilization and FcεRI-mediated serotonin release. This inhibition is mediated by recruitment of the tyrosine phosphatase SHP-1 to the two carboxy-terminal, tyrosine phosphorylated ITIMs in the PIR-B cytoplasmic tail. PIR is expressed on myeloid and B lineage cells in bone marrow; expression levels increase with cellular differentiation and are higher on myeloid cells than on B cells. IL-3-induced bone marrow mast cells also express PIR on the surface. Splenic B cells, macrophages, and dendritic cells express the PIR molecules, while T cells and NK cells do not. The marginal zone B cells express higher levels of PIR than the newly formed and follicular B cells, and the B1 cells also express higher levels of PIR than the B2 cells. Cell surface PIR expression is thus enhanced by cell activation. Given the requirement of FcRγc association for PIR-A cell surface expression, it will be interesting to examine the possibility that the level of FcRγc expression may differentially affect the PIR-A/PIR-B equilibrium as a function of cell lineage and/or activation. While the human gene family ILT/LIR/MIR has 40%–60% homology to the PIR, several questions remain: (1) is the ILT/LIR/MIR gene family the human PIR counterpart?; and (2) are the ligands and functions of the two families the same?

Acknowledgments. We wish to thank our colleagues, Ming Chen, Lanier Gartland, Dong-Wan Kang, Charles Mashburn, and Jin-Yi Wang; our collaborators, Drs. Eric Vivier, Mathieu Bléry, Frédéric Vély, and Jeffrey V. Ravetch for sharing their data, and Marsha Flurry for graphic arts assistance. This work was supported in part by National Institute of Health Grants AI42127, AI39816, and HD36312. MDC is an investigator of the Howard Hughes Medical Institute.

References

Alley TL, Cooper MD, Chen M, Kubagawa H (1998) Genomic structure of PIR-B, the inhibitory member of the paired immunoglobulin-like receptor genes in mice. Tissue Antigens 51:224–231

Arm JP, Nwankwo C, Austen KF (1997) Molecular identification of a novel family of human Ig superfamily members that possess immunoreceptor tyrosine-based inhibition motifs and homology to the mouse gp49B1 inhibitory receptor. J Immunol 159:2342–2349

Bléry M, Kubagawa H, Chen CC, Vély F, Cooper MD, Vivier E (1998) The paired Ig-like receptor PIR-B is an inhibitory receptor that recruits the protein-tyrosine phosphatase SHP-1. Proc Natl acad Sci USA 95:2446–2451

Borges L, Hsu ML, Fanger N, Kubin M, Cosman, D (1997) A family of human lymphoid and myeloid Ig-like receptors, some of which bind to MHC class I molecules. J Immunol 159:5192–5196

Cambier JC (1995) Antigen and Fc receptor signaling: the awesome power of the immunoreceptor tyrosine-based activation motif (ITAM). J Immunol 155:3281–3285

Colonna M, Samaridis J, Cella M, Angman L, Allen RL, O'Callaghan CA, Dunbar R, Ogg GS, Cerundolo V, Rolink A (1998) Human myelomonocytic cells express an inhibitory receptor for classical and nonclassical MHC class I molecules. J Immunol 160:3096–3100

Cosman D, Fanger N, Borges L, Kubin M, Chin W, Peterson L, Hsu ML (1997) A novel immunoglobulin superfamily receptor for cellular and viral MHC class I molecules. Immunity 7:273–282

Daéron M (1997) Fc receptor biology. Annu Rev Immunol 15:203–234

Ernst P, Smale ST (1995) Combinatorial regulation of transcription I: general aspects of transcriptional control. Immunity 2:311–319

Faisst S, Meyer S (1992) Compilation of vertebrate-encoded transcription factors. Nucleic Acids Res 20:3–26

Hayami K, Fukuta D, Nishikawa Y, Yamashita Y, Inui M, Ohyama Y, Hikida M, Ohmori H, Takai T (1997) Molecular cloning of a novel murine cell surface glycoprotein homologous to killer cell inhibitory receptors. J Biol Chem 272:7320–7327

Kubagawa H, Burrows PD, Cooper MD (1997) A novel pair of immunoglobulin-like receptors expressed by B cells and myeloid cells. Proc Natl acad Sci USA 94:5261–5266

Kubagawa H, Chen CC, Hong LH, Shimada T, Gartland L, Mashburn C, Uehara T, Ravetch JV, Cooper MD (1999) Biochemical Nature and Cellular Distribution of the Paired Immunoglobulin-like Receptors, PIR-A and PIR-B 189:309–317

Lanier LL, Corliss BC, Wu J Leong, C, Phillips JH (1998) Immunoreceptor DAP12 bearing a tyrosine-based activation motif is involved in activating NK cells. Nature 391:703–707

Maeda A, Kurosaki M, Ono M, Takai T, Kurosaki T (1998) Requirement of SH2-containing protein tyrosine phosphatases SHP-1 and SHP-2 for paired immunoglobulin-like receptor B (PIR-B)-mediated inhibitory signal. J Exp Med 187:1355–1360

Morton HC, van Egmond M, van de Winkel JGJ (1996) Structure and function of human IgA Fc receptors (FcαR). Crit Rev Immunol 16:423–440

Ravetch JV (1994) Fc receptors: rubor redux. Cell 78:553–560

Reth M (1989) Antigen receptor tail clue. Nature 338:383–384

Samaridis J, Colonna M (1997) Cloning of novel immunoglobulin superfamily receptors expressed on human myeloid and lymphoid cells: structural evidence of new stimulatory and inhibitory pathways. Eur J Immunol 27:660–665

Timms JF, Carlberg K, Gu H, Chen H, Kamatkar S, Nadler MJS, Rohrschneider LR, Nell BG (1998) Identification of major binding proteins and substrates for the SH2-containing protein tyrosine phosphatase SHP-1 in macrophages. Mol Cell Biol 18:3838–3850

Vély F, Vivier E (1997) Conservation of structural features reveals the existence of a large family of inhibitory cell surface receptors and noninhibitory/activatory counterparts. J Immunol 159:2075–2077

Wagtmann N, Rojo S, Eichler E, Mohrenweiser H, Long EO (1997) A new human gene complex encoding the killer cell inhibitory receptors and related monocyte/macrophage receptors. Curr Biol 7:615–618

Weiss A, Schlessinger J (1998) Switching signals on or off by receptor dimerization. Cell 94, 277–280

Yamashita Y, Fukuta D, Tsuji A, Nagabukuro A, Matsuda Y, Nishikawa Y, Ohyama Y, Ohmori H, Ono M, Takai T (1998) Genomic structures and chromosomal location of p91, a novel murine regulatory receptor family. J Biochem (Tokyo) 123:358–368

LAIR-1, a Widely Distributed Human ITIM-Bearing Receptor on Hematopoietic Cells

L. Meyaard

1 Introduction

In the last few decades, immunologists have mainly concentrated on the parameters that involve direct activation of the immune system. However, for the immune system to function properly, fine-tuned negative regulation of an immune response has now proven to be as important as appropriate activation. Failure of inhibitory mechanisms can lead to disastrous phenotypes, as witnessed by the lympho-proliferation in $Fas^{-/-}$, $FasL^{-/-}$, and $CTLA4^{-/-}$ mice; and the multi-focal inflammations in $IL-10^{-/-}$ and $TGF-\beta^{-/-}$ mice. Investigation of the inhibitory mechanisms of the immune system might lead to better insight in the mechanisms underlying autoimmune disease. Here, a novel inhibitory receptor, LAIR-1, is discussed. LAIR-1 has a broad cellular distribution in the immune system and thus might play a role in the regulation of a variety of immune responses.

University Hospital Utrecht, Dept. of Immunology F03.8.21, Heidelberglaan 100, 3584 CX Utrecht, The Netherlands

2 Inhibitory Receptors on Natural Killer Cells

The role of inhibitory receptors has recently been extensively studied on natural killer (NK) cells. The existence of inhibitory NK cell receptors first was suggested by Klas Kärre's "missing-self" hypothesis in 1986 (KÄRRE et al. 1986). The observation that MHC class I deficient lymphomas were selectively rejected in mice led to the hypothesis that NK cells specifically kill tumor cells that lack MHC class I expression. Indeed, in the following years NK cell receptors specific for MHC class I molecules on the target cells were identified and cloned [reviewed in (LANIER 1998)]. In mice, the lectin family Ly-49 binds several different H-2 ligands, whereas the human killer cell inhibitory receptors (KIRs) bind HLA-A, -B, and -C alleles. The KIR-family of molecules, from which about 30 cDNAs have now been identified, belongs to the Ig gene superfamily and is structurally unrelated to Ly49. There is no mouse homologue for KIRs nor a human homologue for Ly49.

In addition to the KIRs, human NK cells express the non-polymorphic receptor CD94/NKG2 A, which binds HLA-E. Upon binding, it transduces a negative signal to the NK cells (BRAUD et al. 1998). Since the non-classical HLA-E molecule is expressed only if complexed with a peptide derived from the leader segment of certain classical HLA-molecules, CD94/NKG2 A is able to keep check on the level of expression of many different MHC class I molecules. Thus, MHC class I receptors prevent killing of autologous cells by NK cells, but permit lysis of virus-infected or tumor cells that have down-regulated MHC class I. Inhibitory MHC class I receptors are also expressed on T cells. KIRs are expressed on small subset of human T cells and have shown to be functional in inhibiting cytotoxicity and cytokine production (D'ANDREA et al. 1996; IKEDA et al. 1997; PHILLIPS et al. 1995).

The KIRs, Ly49, and NKG2 A receptors possess cytoplasmic tails with immune receptor tyrosine-based inhibitory motifs (ITIMs) with the consensus sequence I/VxYxxL/V (THOMAS 1995). These motifs were identified in the inhibitory Fc-receptors on B cells. Simultaneous engagement of the B cell antigen receptor and FcγRIIB1 results in phosphorylation of FcγRIIB1 on tyrosine residues in the ITIM (MUTA et al. 1994), which causes recruitment of SHP-1 (SH2-containing tyrosine phosphatase-1) (D'AMBROSIO et al. 1995). Comparable involvement of the CD22 receptor during BCR stimulation also results in functional inactivation (DOODY et al. 1995). Similarly, upon tyrosine phosphorylation, inhibitory MHC Class I receptors recruit tyrosine phosphatases and thereby inhibit positive signal transduction via other receptors by dephosphorylation of essential components in the positive signaling pathways (FRY et al. 1996; CAMPBELL et al. 1996; BURSHTYN et al. 1996). ITIM-containing receptors might need a simultaneous, positive signal given to the cell to initiate the tyrosine kinase activity and phosphorylate the ITIM.

3 Leukocyte-Associated Immunoglobulin-like Receptors

We have identified and cloned an inhibitory receptor called leukocyte associated immunoglobulin-like receptor (LAIR) (MEYAARD et al. 1997). LAIR-1 is a 32-kDa transmembrane glycoprotein with a single immunoglobulin-like domain. The cytoplasmic domain contains the amino acid sequences VTYAQL and ITYAAV, spaced by 24 aa, which fit the consensus sequences for ITIMs. LAIR-1 is structurally related to the KIRs with 20%–30% homology. The LAIR-1 gene is localized in the same chromosomal region as the KIRs, chromosome 19q13.4. However, LAIR-1 has a much broader cellular distribution than the KIRs and is expressed on the majority of peripheral blood mononuclear cells. In healthy individuals it is expressed on 95%–100% of peripheral blood NK cells and monocytes, on 70%–80% of CD4$^+$ T cells, on 80%–90% of CD8$^+$ T cells, and on 80%–90% of B cells. The majority of human fetal thymocytes also express LAIR-1. Furthermore, it is expressed on human T cell, myeloid, and NK cell tumor cell lines, and all NK and T cell clones. But is not on EBV-transformed B cell lines.

We have recently identified two different cDNAs of the LAIR-family. In addition to LAIR-1, we cloned the gene for LAIR-2, which also contains one Ig-like domain and is 84% homologous to LAIR-1. In contrast to LAIR-1, LAIR-2 lacks a transmembrane and cytoplasmic domain, suggesting that it is a secreted protein.

LAIR-1 functions as an inhibitory receptor on NK cells (MEYAARD et al. 1997). An mAb against LAIR-1 (DX26) inhibits the killing of FcR-bearing targets by both NK cell clones and NK cells freshly isolated from peripheral blood. LAIR-1 needs to be cross-linked to inhibit, as demonstrated by the facts that F(ab')2 fragments have no effects on cytotoxicity and lysis of FcR-negative targets is not affected by DX26 mAb. When NK cell clones are simultaneously activated by mAb against CD16, CD2, CD69, or DNAM-1, DX26 mAb is still able partially or completely to inhibit NK cell-mediated lysis. Thus, cross-linking of LAIR-1 on human NK cells by mAb delivers a potent inhibitory signal capable of inhibiting target cell lysis, mediated by both resting and activated NK cells in vitro, even in the presence of a strong positive signal. Consistent with the inhibitory signal given to NK cells, the cytoplasmic domain of LAIR-1 contains two ITIMs. Indeed, activation of NK cells with sodium pervanadate results in the binding of SHP-1 and SHP-2 to LAIR-1 (MEYAARD et al. 1997).

POGGI et al. reported on a mAb recognizing a molecule of about 40kDa with a similar distribution as LAIR-1, which they designated p40 (POGGI et al. 1995). It and LAIR-1 are the same molecule, as demonstrated by the fact that p40 mAb stains LAIR-1 transfectants (Meyaard et al., unpublished observation). Like DX26 mAb, p40 mAb inhibits NK cell lysis (POGGI et al. 1995, 1997) and the cytotoxicity mediated by some T cell clones (POGGI et al. 1997).

4 LAIR is a Member of the Inhibitory Receptor Superfamily

In the last 2 years, additional genes encoding inhibitory receptors have been recognized and cloned in man and mice [reviewed in (YOKOYAMA 1997)]. Interestingly, all these molecules are related to LAIR and KIR. The inhibitory receptors as a group have recently been indicated as an inhibitory receptor superfamily (LANIER 1998) and are discussed here individually.

In mice, the gp49B molecule (CASTELLS et al. 1994) functions as an inhibitory receptor on mouse mast cells and NK cells (WANG et al. 1997; ROJO et al. 1997). Mouse B cell and myeloid cells express gp91 or paired Ig-like receptors (PIRs) (KUBAGAWA et al. 1997; HAYAMI et al. 1997). Cross-linking of some of these receptors leads to inhibition of BCR-mediated activation of B cells in vitro (BLERY et al. 1998; MAEDA et al. 1998). The ligands of gp49 and PIRs are not known yet, but do not appear to be MHC class I.

Several groups have reported on a novel group of inhibitory molecules that are mainly expressed on human monocytes and B cells. These molecules have been called immunoglobulin-like transcripts (ILTs), leukocyte immunoglobulin-like receptors (LIRs), monocyte/macrophage immunoglobulin-like receptors (MIRs), or HM18 by the laboratories that reported the first cDNA sequences (ARM et al. 1997; WAGTMANN et al. 1997; BORGES et al. 1997; COSMAN et al. 1997; CELLA et al. 1997; SAMARIDIS and COLONNA 1997). Some of the family members have been demonstrated to bind MHC class I molecules or viral MHC class I homologues (COLONNA et al. 1997; BORGES et al. 1997; COSMAN et al. 1997). Most genes are expressed by B cells and monocytic cells, although expression of some family members has been reported on NK and T cells. Cross-linking of the receptors by mAb has been shown to inhibit the antigen-presentation by antigen-presenting cells (CELLA et al. 1997), as well as activation of B cells, T cells, NK cells, and macrophages (COLONNA et al. 1997, 1998). This might indicate that MHC class I also is able to down-regulate function of other cells besides NK cells.

Most of these receptor families contain genes that encode receptors with long cytoplasmic tails that contain ITIMs and function as inhibitory receptors. However, within the families, some genes encode receptors that lack ITIMs and usually have a shorter cytoplasmic tail and a positively charged amino acid residue in their transmembrane region. These molecules are implicated in cellular activation and it has been suggested that they might activate the necessary tyrosine kinases to phosphorylate the ITIMs of their other ITIM containing family members (COLONNA 1998). It has now been shown that, to be able to exert this positive signal, the non-ITIM-containing KIRs, NKG2, and Ly49 receptors associate with a signaling molecule, designated DAP12, which contains ITAMs (immune receptor tyrosine-based activation motifs) responsible for recruitment of tyrosine kinase activity (LANIER et al. 1998a, b; SMITH et al. 1998). Gene families encoding short isoforms have also been reported for the ILTs and PIRs (HAYAMI et al. 1997; SAMARIDIS and COLONNA 1997). So far, we have no indication that similar forms of

LAIR exist, and it seems that the LAIR family is relatively small. LAIR might be exclusively an inhibitory receptor and not paired with an activating counterpart.

5 Future Perspectives

We have not yet identified the ligand(s) for LAIR-1. Since the LAIR-1 mAb does not interfere with HLA-class I recognition by a large panel of NK clones, it seems unlikely that LAIR-1 recognizes HLA class I molecules (MEYAARD et al. 1997). Furthermore, since LAIR-1 is broadly expressed on the majority of mononuclear leukocytes, its ligand(s) may demonstrate a more restricted expression to provide regulation of LAIR function. Identification of the LAIR-ligand(s) will allow us to conclude more on the biological relevance of LAIR, which is as yet only partially known. The inhibitory mechanism of LAIR, using ITIMs, could in principal be active in all cell types. Indeed it has recently been demonstrated that ligation of LAIR on peripheral blood CD14$^+$ cells prevents their maturation into dendritic cells by GM-CSF, indicating that LAIR-1 is functional on these cells (POGGI et al. 1998).

A powerful approach to defining the function of molecules is to create mice in which the gene of interest is disrupted. However, it is possible that LAIR does not have a structural murine homologue, since for other genes located in the same chromosomal region (KIRs, FcRα and ILTs) no murine homologue has been identified to date. However, KIRs do have a functional murine homologue represented by the Ly-49 family. Identification of LAIR-ligand might elucidate whether there is a functional homologue in the mouse.

Acknowledgments. Most of this work was performed in the Department of Immunobiology of the DNAX Research Institute in Palo Alto, California, which is supported by the Schering Plough Cooperation. The project is being continued in the Department of Immunology of the University Hospital in Utrecht, The Netherlands. L.M. is supported by the Netherlands Organization of Scientific Research (NWO). L. Lanier, J. Phillips, and H. Clevers are acknowledged for critical reading of the manuscript.

References

Arm JP, Nwankwo C, Austen KF (1997) Molecular identification of a novel family of human Ig superfamily members that possess immunoreceptor tyrosine-based inhibition motifs and homology to the mouse gp49B1 inhibitory receptor. J Immunol 159:2342–2349

Blery M, Kubagawa H, Chen C-C, Vely F, Cooper MD, Vivier E (1998) The paired Ig-like receptor PIR-B is an inhibitory receptor that recruits the protein-tyrosine phosphatase SHP-1. Proc Natl Acad Sci USA 95:2446–2451

Borges L, Hsu M-L, Fanger N, Kubin M, Cosman D (1997) A family of human lymphoid and myeloid Ig-like receptors, some of which bind to MHC Class I molecules. J Immunol 159:5192–5196

Braud VM, Allan DSJ, O'Callaghan CA, Soderstrom K, D'Andrea A, Ogg GS, Lazetic S, Young NT, Bell JI, Phillips JH, Lanier LL, McMichael AJ (1998) HLA-E binds to natural killer cell receptors CD94/NKG2 A,B, and C. Nature 391:795–799

Burshtyn DN, Scharenberg AM, Wagtmann N, Rajagopolan S, Berrada K, Yi T, Kinet J-P, Long EO (1996) Recruitment of tyrosine phosphatase HCP by the killer cell inhibitory receptor. Immunity 4:77–85

Campbell KS, Dessing M, Lopez-Botet M, Cella M, Colonna M (1996) Tyrosine phosphorylation of a human killer inhibitory receptor recruits protein tyrosine phosphatase 1C. J Exp Med 184:93–100

Castells MC, Wu X, Arm JP, Austen KF, Katz HR (1994) Cloning of the gp49B gene of the immunoglobulin superfamily and demonstration that one of its two products is an early expressed mast cell surface protein originally described as gp49. J Biol Chem 269:8393–8401

Cella M, Dohring C, Samaridis J, Dessing M, Brockhaus M, Lanzavecchia A, Colonna M (1997) A novel inhibitory receptor (ILT3) expressed on monocytes, macrophages, and dendritic cells involved in antigen processing. J Exp Med 185:1743–1751

Colonna M, Navarro F, Bellon T, Llano M, Garcia P, Samaridis J, Angman L, Cella M, Lopez-Botet M (1997) A common inhibitory receptor for major histocompatibility complex class I molecules on human lymphoid and myelomonocytic cells. J Exp Med 186:1809–1818

Colonna M (1998) Unmasking the killer's accomplice. Nature 391:642–643

Colonna M, Samaridis J, Cella M, Angman L, Allen RL, O'Callaghan CA, Dunbar R, Ogg GS, Cerundolo V, Rolink A (1998) Human myelomonocytic cells express an inhibitory receptor for classical and nonclassical MHC class I molecules. J Immunol 160:3096–3100

Cosman D, Fanger N, Borges L, Kubin M, Chin W, Peterson L, Hsu M-L (1997) A novel immunoglobulin superfamily receptor for cellular and viral MHC class I molecules. Immunity 7:273–282

D'Ambrosio D, Hippen KL, Minskoff SA, Mellman I, Pani G, Siminovitch KA, Cambier JC (1995) Recruitment and activation of PTP1C in negative regulation of antigen receptor signaling by FcgRIIB1. Science 268:293–297

D'Andrea A, Chang C, Phillips JH, Lanier LL (1996) Regulation of T cell lymphokine production by killer cell inhibitory receptor recognition of self HLA class I alleles. J Exp Med 184:789–794

Doody GM, Justement LB, Delibrias CC, Matthews RJ, Lin J, Thomas ML, Fearon DT (1995) A role in B cell activation for CD22 and the protein tyrosine phosphatase SHP. Science 269:242–244

Fry AM, Lanier LL, Weiss A (1996) Phosphotyrosines in the killer cell inhibitory receptor motif of NKB1 are required for negative signaling and for association with protein tyrosine phosphatase 1C. J Exp Med 184:295–300

Hayami K, Fukuta D, Nishikawa Y, Yamashita Y, Inui M, Ohyama Y, Hikida M, Ohmori H, Takai T (1997) Molecular cloning of a novel murine cell-surface glycoprotein homologous to killer cell inhibitory receptors. J Biol Chem 272:7320–7327

Ikeda H, Lethe B, Lehmann F, Van Baren N, Baurain J-F, De Smet C, Chambost H, Vitale M, Moretta A, Boon T, Coulie PG (1997) Characterization of an antigen that is recognized on a melanoma showing partial HLA loss by CTL expressing an NK inhibitory receptor. Immunity 6:166–208

Kärre K, Ljunggren HG, Piontek G, Kiessling R (1986) Selective rejection of H-2-deficient lymphoma variants suggests an alternative immune defence strategy. Nature 319:675–678

Kubagawa H, Burrows PD, Cooper MD (1997) A novel pair of immunoglobulin-like receptors expressed by B cells and myeloid cells. Proc Natl Acad Sci USA 94:5261–5266

Lanier LL (1998) NK cell receptors. Annu Rev Immunol 16:359–393

Lanier LL, Corliss B, Wu J, Phillips JH (1998a) Association of DAP12 with activating CD94/NKG2 C NK cell receptors. Immunity 8:693–701

Lanier LL, Corliss BC, Wu J, Leong C, Phillips JH (1998b) Immunoreceptor DAP12 bearing a tyrosine based activation motif is involved in activating NK cells. Nature 391:703–707

Maeda A, Kurosaki M, Ono M, Takai T, Kurosaki T (1998) Requirement of SH2-containing protein tyrosine phosphatases SHP-1 and SHP-2 for paired immunoglobulin-like receptor B (PIR-B)-mediated inhibitory signal. J Exp Med 187:1355–1360

Meyaard L, Adema GJ, Chang C, Woollatt E, Sutherland GR, Lanier LL, Phillips JH (1997) LAIR-1, a novel inhibitory receptor expressed on human mononuclear leukocytes. Immunity 7:283–290

Muta T, Kurosaki T, Misulovin Z, Sanchez M, Nussenzweig MC, Ravetch JV (1994) A 13-amino-acid motif in the cytoplasmic domain of FcgammaRIIB modulates B-cell receptor signaling. Nature 368:70–73

Phillips JH, Gumperz JE, Parham P, Lanier LL (1995) Superantigen-dependent, cell-mediated cytotoxicity inhibited by MHC Class I receptors on T lymphocytes. Science 268:403–405

Poggi A, Pella N, Morelli L, Spada F, Revello V, Sivori S, Augugliaro R, Moretta L, Moretta A (1995) p40, a novel surface molecule involved in the regulation of the non-major-histocompatibility-complex-restricted cytolytic activity in humans. Eur J Immunol 25:369–376

Poggi A, Tomasello E, Revello V, Nanni L, Costa C, Moretta L (1997) p40 molecule regulates NK cell activation mediated by NK receptors for HLA class I antigens and TCR-mediated triggering of T lymphocytes. Int Immunol 9:1271–1279

Poggi A, Tomasello E, Ferrero E, Zocchi MR, Moretta L (1998) p40/LAIR-1 regulates the differentiation of peripheral blood precursors to dendritic cells induced by granulocyte-monocyte colony-stimulating factor. Eur J Immunol 28:2086–2091

Rojo S, Burshtyn DN, Long EO, Wagtmann N (1997) Type I transmembrane receptor with inhibitory function in mouse mast cells and NK cells. J Immunol 158:9–12

Samaridis J, Colonna M (1997) Cloning of novel immunoglobulin superfamily receptors expressed on human myeloid and lymphoid cells: structural evidence for new stimulatory and inhibitory pathways. Eur J Immunol 27:660–665

Smith KM, Wu J, Bakker ABH, Phillips JH, Lanier LL (1998) Ly-49D and Ly-49H associate with mouse DAP12 and form activating receptors. J Immunol 161:7–10

Thomas ML (1995) Of ITAMs and ITIMs: turning on and off the B cell antigen receptor. J Exp Med 181:1953–1956

Wagtmann N, Rojo S, Eichler E, Mohrenweiser H, Long EO (1997) A new human gene complex encoding the killer cell inhibitory receptors and related monocyte/macrophage receptors. Curr Biol 7:615–618

Wang LL, Mehta IK, LeBlanc PA, Yokoyama WM (1997) Mouse natural killer cells express gp49B1, a structural homologue of human killer inhibitory receptors. J Immunol 158:13–17

Yokoyama WM (1997) What goes up must come down: The emerging spectrum of inhibitory receptors. J Exp Med 186:1803–1808

The Mast Cell Function-Associated Antigen, a New Member of the ITIM Family

R. Xu and I. Pecht

1 Introduction

A decade has already passed since we discovered, essentially by serendipity, the membranal glycoprotein later named MAst cell Function-associated Antigen (MAFA). While trying to identify the still elusive Ca^{2+} channel responding to the type 1 Fcε receptor (FcεRI) stimulus, a series of monoclonal antibodies (mAb) were raised to the rat mucosal-type mast cells of the RBL-2H3 line. One of these mAbs, G63, caused a dose-dependent inhibition of the secretory response to the FcεRI stimulus and was shown to bind to a membranal protein which we proceeded to isolate and characterize (ORTEGA-SOTO and PECHT 1988).

Department of Immunology, The Weizmann Institute of Science, Rehovot 76100, Israel

2 Mast Cell Function-Associated Antigen, Encoding Genes and Proteins

The amino acid sequence deduced from the cloned, full-length cDNA showed that it is a type II membrane glycoprotein (GUTHMANN et al. 1995a). The extracellular C-terminal 114- amino acids of MAFA has a marked homology with the carbohydrate recognition domain (CRD) of several members of the C-type (calcium-dependent) animal lectin family, e.g., type II FcεR (CD23), CD69, CD72, NKR-P1, or Ly49. It's sequence includes 15 conserved residues (six cysteines, five tryptophans, two glycines, and two leucines) interspersed within this 114- to 129-amino acid-long CRD, as well as the characteristic WIGL and CYYF motifs. The cytoplasmic tail of MAFA contains an SIYSTL motif, clearly related to the V/IX-YXXL sequence recently named immunoreceptor tyrosine-based inhibitory motif (ITIM; AMIGORENA et al. 1992). A similar motif has also been found among several of the related C-type lectins (GUTHMANN et al. 1995a).

Rat MAFA was found to be encoded by a single-copy gene that spans 13kb in the rat genome and is composed of five exons separated by four introns, similar to the mouse and human CD69. Three separate exons encode the CRD of MAFA. Functional analysis of the 5' flanking region of the gene revealed that a cell type-specific promoter is located within the first 664bp upstream of the transcription origin. The promoter lacks any obvious TATA box and drives gene transcription originating from multiple start sites. Examination of the MAFA transcripts for possible polymorphism revealed the presence of two novel ones, generated by alternative splicing; deletion of the exon encoding the transmembranal domain in one of them does not cause a frame shift and would, upon translation, give rise to a soluble MAFA molecule. This suggests that, like FcεRII, FcγRII, and CD72, the MAFA may also exist as a soluble factor and not only in a membrane-bound form. Splicing of two exons in a second transcript results in a new reading frame encoding a putative protein which contains MAFA's cytoplasmic domain (BOCEK et al. 1997).

MAFA has a high degree of glycosylation: while its polypeptide core amounts to about 19kDa, N-linked oligosaccharides account for up to half of the apparent molecular mass, which yields species with up to ~40kDa. The protein originally isolated by immunoprecipitation with mAb G63 from the RBL-2H3 cells exhibited two rather broad bands (28–40 and 60–82kDa) in non-reducing sodium dodecyl sulfate polyacrylamide gel electrophoresis (SDS-PAGE). These two bands coalesced upon reduction to the lower molecular mass band (ORTEGA SOTO and PECHT 1988). Peptide mapping has established that the above two species are due to the monomeric and disulfide-linked homodimeric forms of the MAFA, respectively (GUTHMANN et al. 1995a).

Slight differences, most probably caused by different glycosylation, were found in the properties of the expressed MAFA when its gene was transfected into different cell lines; When MAFA cDNA including the second ATG codon was expressed in COS cells, a single band of 58kDa was observed in nonreducing SDS-PAGE and 28kDa under reducing conditions (Bocek and Pecht, unpublished).

When it was expressed in insect cells by the baculovirus expression system (BES), both monomeric and homodimeric recombinant MAFA (rMAFA) were of slightly smaller apparent molecular masses than the respective molecular species expressed by the RBL-2H3 cells. Both MAFA species expressed either in COS or insect cells were identified by binding to the mAb G63 (BINSACK and PECHT 1997).

More recently, the gene encoding a human MAFA analog was cloned and sequenced from the human basophil-like leukemia cell line KU812 as well as from human lung mast cells. The deduced amino acid sequence of the human MAFA has 51% homology with that of the rat. It also possesses an intracellular ITIM-like motif with the sequence VIYSML (rat is SIYSTL) and an extracellular C-lectin-like CRD (BUTCHER et al. 1998). The mouse MAFA encoding cDNA has also been cloned recently and, like the rat MAFA, it is also encoded by a single gene. An ITIM motif was also present in the cytoplasmic tail of mouse MAFA with an identical sequence (SIYSTL; BLASER et al. 1998). The high degree of sequence conservation between the three species suggests that MAFA has an important and evolutionary conserved function.

3 Monomer/Dimer Forms of the MAFA

The highly conserved cysteine residues in the extracellular domain (ECD) suggested an important role for the intrachain disulfide bonds in attaining its fold and biological function. In order to identify the cysteine residues involved in the interchain disulfide linkage, site-directed mutagenesis of MAFA cDNA based on polymerase chain reaction (PCR) has been performed, converting several cysteines to serine residues. The PCR products carrying the specific substitutions were subsequently subcloned into eukaryotic expression vectors and transiently expressed in COS-7 cells. The presence, in these cells, of the recombinant MAFA protein in dimeric or monomeric forms was analyzed by immunoprecipitation and SDS-PAGE. The only cysteinyl residue present in the cytosolic domain of MAFA (Cys 34) was considered the most likely one to be involved in the process of the MAFA dimer formation (BOCEK 1996). However, mutation of the two juxtamembranal cysteinyl residues in the ECD (Cys61 and Cys62) in combination with this cytosolic cysteinyl reduced the amount of the MAFA dimer in favor of the monomer by only ~30%. Thus, the identity of the residues responsible for the formation of MAFA homodimers is still unresolved.

On the functional level, it is still unclear whether formation of the MAFA homodimer is essential for its activity: whereas G63-Affigel immunoprecipitates both monomeric and dimeric MAFA from cell lysates, soluble mAb G63 reacted with the intact cells prior to lysis led to immunoprecipitation of dimeric MAFA with only a trace of the monomer after the addition of Protein-A. Adding increasing concentrations of soluble mAb G63 to the cell lysates, immunoprecipitated the monomeric MAFA as well (Xu and Pecht, unpublished). These results most probably reflect the expected differences in avidity between the monomeric and the dimeric MAFA for mAb G63. It was already established very early that MAFA's inhibitory capacity of the FcεRI-induced secretory response requires its clustering, as the monovalent Fab fragments of G63 do not cause any inhibition, whereas the intact mAb and its (Fab')$_2$ fragment do (ORTEGA SOTO and PECHT 1988; Xu and Pecht, unpublished). Thus, only the clustered form of the MAFA yields a functional species for recruiting and activating the kinases or phosphatases responsible for initiation of its signaling cascades (see below). However, so far we do not know to what extent the monomeric and homodimeric MAFA differ functionally.

4 Tissue and Cell Distribution

By binding the radio-iodinated Fab fragment of mAb G63, the number of MAFA molecules per RBL-2H3 cell was determined to be about 2×10^4/cell. This is 20–30 times lower than that of FcεRI on the same cells (ORTEGA SOTO and PECHT 1988). Expression of the MAFA gene or protein by other cell types was pursued by using both RT-PCR and mAb G63. In RT-PCR, two distinct oligonucleotide probes were employed; one for both, the full length MAFA transcripts and those from which exon 2 has been spliced out, and the other for only the full-length MAFA transcripts. Screening of rat lung, muscle, kidney, testis, Peyer's patch, spleen, liver, brain, and intestine, by that method detected transcription of the MAFA gene in lung only (BOCEK et al. 1997). Two distinct amplified DNA fragments were observed in both lung tissue and RBL-2H3 cells, suggesting that, in addition to the full-length MAFA, that from which exon 2 was spliced out is also present. However, the intensity of the expression detected in lungs was markedly lower than in the RBL-2H3 cells, suggesting that it is limited to mast cells only. Immunohistochemical analysis carried out using mAb G63 in frozen lung sections has indeed shown staining of mast cells (BOCEK et al. 1997). Earlier, independent-flow cytometric measurements showed mAb G63 binding to freshly isolated human basophils (GELLER-BERNSTEIN et al. 1994). Taken together, these results suggest that MAFA expression is limited to specific organs, in accordance with its being, at least in the rat, a mast-cell-specific molecule (BOCEK et al. 1997).

Except on mast cells, human MAFA was also found to be expressed by peripheral blood natural killer cells and by the monocyte-like cell line U937 (BUTCHER et al. 1998). Unlike rat MAFA, mouse MAFA was expressed in effector CD8 T-cells and by lymphokine-activated NK cells, but not on mast cells (BLASER et al. 1998). Once more, MAFA expression by multiple types of immunocytes indicates that it may have an important role in host defense.

5 Relation to FcεRI

Early fluorescence microscopy studies on RBL-2H3 cells have shown that clustering of FcεRI-IgE on RBL-2H3 cells by multivalent antigen results in redistribution of MAFA, leading to its co-localization with the aggregated FcεRI-IgE (ORTEGA et al. 1991). This association of MAFA with the FcεRI does not involve cytoskeletal elements, as it was unaffected by reagents such as cytochalasin. Moreover, FcεRI aggregation also promoted MAFA's internalization, and the time course of antigen-induced internalization of FcεRI and MAFA were found to be similar, which further suggested a direct interaction between them.

Indeed, studies of fluorescence resonance energy transfer (FRET) between MAFA-bound, fluorophore labeled mAb G63 and FcεRI-bound labeled IgE provided further evidence for such proximity. However, no difference was observed between results of FRET measurements on resting cells and those on which FcεRI-IgE clustering was caused by a polyvalent antigen (JURGENS et al. 1996). As these measurements were done using the intact mAb G63(IgG1), the possibility that MAFA clustering may have caused this proximity cannot be excluded. Thus, proximity between the FcεRI and clustered MAFA probably preexists on resting cells, and clustering of either of these components may further promote the association. Further FRET studies aimed at resolving this issue are in progress. Studies of RBL-2H3's secretion using noncovalent hetero- and homodimers of Fab fragments of G63 and J17 (a mAb specific for the FcεRI) were recently carried out to examine possible functional reflection of this association. A quantitative analysis of the secretory response to these reagents provided evidence that ∼60% of MAFA molecules may associate with the FcεR1 on the RBL-2H3 cells (SCHWEITZER-STENNER et al. 1999).

6 Identification of Ligands

The homology between the C-terminal 114-amino acid sequence of the ECD of the MAFA and the CRD of other C-type lectins strongly suggested its carbohydrate binding capacity. In order to search for potential MAFA ligands, recombinant MAFA (rMAFA) was produced using the BES. The rMAFA was expressed using the *Spodoptera frugiperda* insect cell line (Sf9) and a recombinant baculovirus. Using a series of different neoglycans, rMAFA was indeed shown to bind to those bearing α-D-mannose residues, suggesting that it has an affinity for terminal mannose residues. Detailed, comparison showed that its saccharide binding pattern is very similar to that of the plant lectin concanavalin A. Furthermore, mannose-carrying neoglycan binding could be inhibited by an excess of α-D-mannose or by EGTA, indicating that the MAFA-saccharide binding is reversible and Ca^{2+}-dependent, as expected from a C-type lectin (BINSACK and PECHT 1997). These results still leave open the question of the physiologically specific MAFA ligand (e.g., is it a glycolipid or a glycoprotein and what is its identity?). In addition, they do not exclude the possibility that MAFA binds to a protein epitope of a ligand, as has been found for MAFA homologues such as CD23.

7 Functional Role

7.1 Significance of Its Carbohydrate Binding

As mentioned above, MAFA clustering by mAb G63 causes a dose-dependent inhibition of mast cells' (such as the RBL-2H3 and mouse bone marrow-derived mast cells') secretory response to FcεRI aggregation. Inhibition is also caused to the de novo synthesis and secretion of inflammatory mediators such as arachidonic acid and LTC4, as well as of cytokines such as tumor necrosis factor α (TNFα) and interleukin 6 (IL6) (GUTHMANN et al. 1995b). At a low degree of FcεRI aggregation, almost complete inhibition of the release of stored mediators can be achieved, whereas at maximal stimulation the inhibition reaches only 30%–40%. A clear correlation was established between mAb G63 binding and its inhibitory effect (ORTEGA SOTO and PECHT 1988).

As expected from MAFA's carbohydrate-binding capacity, evidence was found for its possible involvement in mast cell adhesion. Using Northern blot analysis, it was observed that, during RBL-2H3 cell adhesion, a marked upregulation of MAFA transcription takes place (BOCEK 1996). The expression of MAFA-specific mRNA was already enhanced 3h after plating the cells, and reached its maximum between 6 and 12h, followed by a decline. Flow cytometric analysis of MAFA expression using biotinylated G63 and Phycoerythrin-Streptavidin has shown that a significant increase in MAFA expression took place during the adhesion process of adherent as well as nonadherent subpopulations of RBL-2H3 cells. Thus, MAFA membrane expression is rapidly induced in these cells during their adhesion (COHEN-DAYAG et al. 1992). Furthermore, a reduction in MAFA expression by RBL-2H3 cells was caused by introducing antisense oligonucleotides complementary to the MAFA mRNA, which was paralleled by a decrease in their adhesion (GUTHMANN 1994; GUTHMANN et al., 1995b). A significant decrease was also observed in RBL-2H3 cell adhesion in the presence of intact G63 or its Fab or F(ab')$_2$ fragments (BOCEK 1996).

Taken together, these various experimental results suggest the existence of a positive correlation between RBL-2H3 cells' adhesion and MAFA expression on its surface. Further studies will establish whether and how this is related to the function of the MAFA and to its lectin structure.

7.2 Mechanism(s) of MAFA Inhibitory Activity

7.2.1 What Cellular Events Does MAFA Clustering Cause?

The question at which step of the FcεRI stimulus-response coupling cascade MAFA interferes has been addressed already, following its discovery using mAb G63. Detailed binding measurements of IgE and mAb G63 to RBL-2H3 cells clearly established that there is no interference of either ligand with the binding of the other (ORTEGA SOTO and PECHT 1988). Later on, it was investigated whether clustering MAFA itself causes any cellular changes and, surprisingly, a time-dependent enhancement of the tyrosyl phosphorylation of the β-subunit of FcεRI was observed, while no change was detected in the FcεRI-γ chain (XU and PECHT 1996). It has already been amply documented that the src family protein tyrosine kinase (PTK) Lyn associates with the FcεRI-β chain already in resting RBL-2H3 cells (EISEMAN and BOLEN 1992; PAOLINI et al. 1992). In response to the FcεRI clustering, Lyn phosphorylates the FcεRI-β chain, thereby initiating the coupling cascade. Immunoprecipitation of FcεRI-β subunit from resting and MAFA-clustered RBL-2H3 cell lysates, showed that the latter enhances Lyn binding to FcεRI-β and apparently leads to the increase in its tyrosyl phosphorylation (Xu and Pecht, 1996 and unpublished). These results constitute the first evidence for the capacity of clustered MAFA to produce, on its own, biochemical changes in mast cells that are most probably related to its inhibitory action.

7.2.2 Interference with the FcεRI Signaling Cascade

One important feature that characterizes the inhibitory action of MAFA is that, unlike other ITIM containing receptors, its inhibition of the FcεRI secretory response does not require their co-clustering. The reason for this might be the observed proximity and association between MAFA and FcεRI. Interestingly, a recent study of the inhibition of antigen-induced T-cell response and antibody-induced NK cell cytotoxicity by the CD94-NKG2A heterodimer was also attained following its self-clustering (LE DRÉAN et al. 1998).

mAb G63's capacity to suppress early events in the FcεRI stimulus-response coupling cascade was established already in the first phase of characterizing its properties (ORTEGA SOTO and PECHT 1988). Thus, hydrolysis of phosphatidylphosphoinositides and the Ca^{2+} signal were suppressed upon pre-treating FcεRI stimulated (but not Ca^{+2} ionophore treated) RBL-2H3 cells with mAb G63 or its $F(ab')_2$ fragments. These results have indicated that the interference takes place upstream to the activation of PLCγ (SCHNEIDER et al. 1992). Recently, it was further shown that MAFA clustering inhibits the antigen-induced tyrosyl phosphorylation and activation of Syk. Moreover, MAFA clustering was also found to enhance the recruitment of SHP-2 to Syk. In vitro experiments showed that SHP-2 isolated from Ag-stimulated RBL-2H3 cells dephosphorylates the FcεRI-induced phosphorylated tyrosine(s) of Syk. As MAFA clustering was also shown earlier to enhance tyrosyl phosphorylation of Lyn, and Lyn was shown

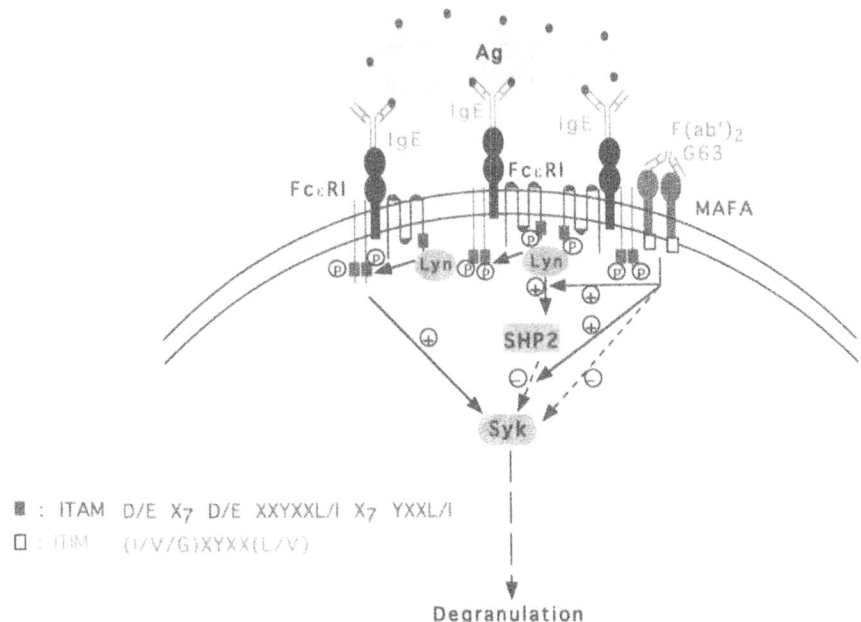

Fig. 1. Some features of MAFA's interference with the FcεRI stimulus. *Arrows* with " + " indicate an activatory process, those with "–" an inhibitory one

to bind to the SH2 domains of SHP-2 (Xu and Pecht, unpublished), a tentative, indirect pathway for the inhibitory action of MAFA may be outlined. Its clustering enhances tyrosyl phosphorylation of Lyn and thereby Lyn association with SHP-2 (via the latter's SH2 domains). This causes SHP-2 activation, which dephosphorylates and thereby suppresses Syk activity (Fig. 1). These results are in line with the well-established view of the crucial role of Syk activation in the FcεRI signaling cascade (TAYLOR et al. 1995; ZHANG et al. 1996).

7.2.3 MAFA's Cytosolic Tail Behaves as an ITIM Containing Element

The functional role of the cytosolic tail of MAFA and identification of potential intracellular proteins that interact with it were and are being pursued by several approaches. First, peptides with the sequence corresponding to residues 4–12 of MAFA's N-terminal tail (MAFA ITIM) were synthesized also with their phosphorylated or thiophosphorylated tyrosyl and used in affinity chromatography of RBL-2H3 cell lysates. It was found that the latter two peptides bound all three phosphatases: SHP-1, SHP-2, and SHIP. The nonphosphorylated peptide or, the control, reversed sequence peptide, did not bind any of these. Second, those results were corroborated by surface plasmon resonance (SPR) analysis using recombinant soluble tandem SH2 domains of SHP1, SHP2, SHIP, and biotin-labeled and tyrosyl phosphorylated MAFA ITIM peptide, showing that this association is a direct one (Philosof-Oppenheimer, R. unpublished).

Taken together, these results clearly suggest that the SIYSTL motif present in MAFA's cytosolic tail behaves as an ITIM. Furthermore, it shows that the amino acid in position Y-2 of the canonical ITIM motif (I/VXYXXL) is not restricted to the hydrophobic isoleucine or valine, as a serine residue at this position also maintains the binding capacity of all the above three phosphatases. It therefore suggests that MAFA is a legitimate new member of the growing ITIM-containing protein family.

Acknowledgments. Studies reported in this paper were supported by grants from the Israel National Science Foundation and the Minerva Foundation, Munich, Germany.

References

Amigorena S, Bonnerot C, Drake J, Choquet D, Hunziker W, Guillet JG, Webster P, Sautes C, Mellman I, Fridman WH (1992) Cytoplasmic domain heterogeneity and functions of IgG Fc receptors in B-lymphocytes. Science 256:1808–1812

Binsack R, Pecht I (1997) The mast cell function-associated antigen exhibits saccharide binding capacity. Eur J Immunol 27:2557–2561

Blaser C, Kaufmann M, Pircher H (1998) Virus-activited CD8 T cells and lymphokine-activated NK cells express the mast cell function-associated antigen MAFA an inhibitory C-type lectin. J Immunol 161:6451–6454

Bocek PJ, Guthmann MD, Pecht I (1997) Analysis of the genes encoding the mast cell function-associated antigen and its alternatively spliced transcripts. J Immunol 158:3235–3243

Bocek PJ (1996) Ph.D. Thesis, Feinberg Graduate School, Rehovot, ISRAEL

Butcher S, Arney KL, Cook GP (1998) MAFA-L an ITIM-containing receptor encoded by the human NK cell gene complex and expressed by basophils and NK cells. Eur J Immunol 28:1–8

Cohen-Dayag A, Schneider H, Pecht I (1992) Variants of the mucosal mast cell line (RBL-2H3) deficient in a functional membrane glycoprotein. Immunobiology 185:124–149

Eiseman E, Bolen JB (1992) Engagement of the high-affinity IgE receptor activates src protein-related tyrosine kinases. Nature (Lond.) 355:78–80

Geller-Bernstein C, Berrebi A, Bassous Gedj L, Ortega E, Licht A, Pecht I (1994) Antibodies specific to membrane components of rat mast cells cross-react with human basophils. Int Arch Allergy Immunol 105:269–273

Guthmann MD (1994) Ph.D. Thesis, Feinberg Graduate School, Rehovot, ISRAEL

Guthmann MD, Tal M, Pecht I, (1995a) A secretion inhibitory signal transduction molecule on mast cells is another C-type lectin. Proc Natl Acad Sci USA 92:9397–9401

Guthmann MD, Tal M, Pecht I, (1995b) A new member of the C-type lectin family is a modulator of the mast cell secretory response. Int Arch Allergy Immunol 107:82

Jurgens L, Arndt-Jovin D, Pecht I, Jovin TM (1996) Proximity relationships between the type I receptor for Fcε (FcεRI) and the mast cell function-associated antigen (MAFA) studied by donor photo-bleaching fluorescence resonance energy transfer microscopy. Eur J Immunol 26:84–91

Le Dréan E, Vély F, Olcese L, Cambiaggi A, Guia S, Krystal G, Gervois N, Morreta A, Jotereau F, Vivier E, (1998) Inhibition of antigen-induced T cell response and antibody-induced NK cell cyto-toxicity by NKG2: A association of NKG2A with SHP-1 and SHP-2 protein tyrosine phosphatases. Eur J Immunol 28:264–276

Ortega Soto E, Pecht I (1988) A monoclonal antibody that inhibits secretion from rat basophilic leukemia cells and binds to a novel membrane component. J Immunol 141:4324–4332

Ortega E, Schneider H, Pecht I, (1991) Possible interactions between the Fcε receptor and a novel mast cell function-associated antigen. Int Immunol 3:333–342

Paolini R, Numerof R, Kinet JP (1992) Phosphorylation/dephosphorylation of high-affinity IgE receptors- a mechanism for coupling a large signaling complex. Proc Natl Acad Sci USA 89:10733–10737

Schneider H, Cohen-Dayag A, Pecht I (1992) Tyrosine phosphorylation of phospholipase Cγ1 couples the Fcϵ receptor mediated signal to mast cells secretion. Internatl Immunol 4:447–453

Schweitzer-Stenner R, Engelke M, Licht A, Pecht I (1999) Mast cell stimulation by co-clustering the type I Fcϵ-receptors with mast cell function-associated antigens. Immunol Letters 68:00–00

Taylor JA, Karas JL, Ram MK, Green OM, Seidel-Dugan C (1995) Activation of the high affinity immunoglobulin E receptor FcϵRI in RBL-2H3 cells is inhibited by Syk SH2 domains. Mol Cell Biol 15:4149–4157

Xu R, Pecht I (1996) Clustering the mast cell function-associated antigen(MAFA) induces tyrosyl phosphorylation of the FcϵR1-β subunit. Immunol Letters 54:105–108

Zhang J, Berenstein EH, Evans RL, Siraganian RP (1996) Transfection of Syk protein tyrosine kinase reconstitues high affinity IgE receptor-mediated degranulation in a Syk-negative variant of rat basophilic leukemia RBL-2H3 cells. J Exp Med 184:71–80

The Enigma of Activating Isoforms of ITIM-Bearing Molecules

A. Cambiaggi[1], M. Lucas[2], E. Vivier[2,3], and F. Vély[2]

1 Introduction

A novel family of inhibitory receptors has been described recently (Vivier and Daeron 1997) based on the presence of a conserved immunoreceptor tyrosine-based inhibition motif (ITIM) in the intracytoplasmic portion. The first ITIM-bearing receptors were identified, in cells of hematopoietic origin, as the low affinity receptors for immunoglobulins, FcγRIIB, in B cells (Amigorena et al. 1992; Daëron et al. 1995) and the killer cell Ig-like receptors (KIR) for MHC class I molecules in natural killer (NK) cells (Olcese et al. 1996; Burshtyn et al. 1996). Subsequently, other ITIM-bearing molecules have been described in both hematopoietic cells [PIR-B, (Kubagawa et al. 1997), ILT-2 and -3, (Cella et al. 1997; Samaridis and Colonna 1997), LIR molecules (Borges et al. 1997), MIR molecules, (Wagtmann et al. 1997), CD22, (Sato et al. 1998), and CD72, (Wu et al. 1998)] and nonhematopoietic cells [SIRPα, (Kharitonenkov et al. 1997)]. All these molecules are characterized by the presence of one or more ITIM sequences in their intracytoplasmic portions. This motif is responsible for the transduction of inhibitory signals inside the cells. The ITIM-bearing receptors belong to two families: the immunoglobulin superfamily (IgSF) and the lectin-like superfamily (LSF).

[1] Unité de Biologie Moléculaire du Gène, Institut Pasteur, 25 rue Dr. Roux, F-75015 Paris, France
[2] Centre d'Immunologie INSERM-CNRS de Marseille-Luminy, Case 906, F-13288 Marseille cedex 09, France
[3] Institut Universitaire de France

Strikingly, activating counterparts of the ITIM-bearing receptors have been characterized by high homology (around 90%) in the extracellular portion, as well as major differences in the transmembrane and intracytoplasmic portions: the presence of a charged amino acid residue in the transmembrane domain and lack of the ITIM motif in the intracytoplasmic domain (VELY and VIVIER 1997).

2 Human and Mouse Activating NKR

Among the ITIM-bearing receptors, the natural killer cell receptors (NKR) for MHC class I molecules are well characterized in both humans and mice (LANIER 1998a).

NK cells are able to recognize MHC class I molecules through specific inhibitory NKR. The IgSF NKR are defined as killer-cell Ig-like receptors (KIR). They belong to a multigenic and multiallelic superfamily and all the genes are localized on the chromosome 19q13.4. KIR are characterized by two or three immunoglobulin-like domains in their extracytoplasmic portion, a non-polar transmembrane portion, and a long intracellular domain that contains two ITIM (except for KIR2DL4, which contains only one ITIM) (Fig. 1). The integrity of ITIM sequences is necessary for the recruitment of intracytoplasmic effector molecules involved in the inhibition of NK cell activity (i.e., the protein tyrosine phosphatases SHP-1 and SHP-2). In humans, the lectin-like NKR are formed of heterodimers between the CD94 molecule and members of the NKG2 family. In mice, NKR are constituted by homodimers of C-type lectins belonging to the Ly49 family (TAKEI et al. 1997); and recently the mouse CD94/NKG2 A molecules have also been described (VANCE et al. 1998).

The first activating KIR molecules to be identified were the counterparts of KIR specific for HLA-C alleles. The same monoclonal antibodies (mAb) reacting with the inhibitory forms also recognize molecules capable of triggering NK cell activation functions as judged by Ca^{2+} mobilization, cytolytic activity, and cytokine production (MORETTA et al. 1995). The cloning of KIR2DS1 and KIR2DS2 revealed a high homology (up to 96%) with their inhibitory counterparts in the extracellular domain, the presence of a lysine in the transmembrane domain, and the lack of the ITIM regions in the intracytoplasmic domain due to an in-frame stop codon within the first ITIM (Fig. 1) (BIASSONI et al. 1996). Despite the high degree of homology with the inhibitory KIR in the extracytoplasmic portion, the HLA specificity of KIR2DS1 and KIR2DS2 is not completely unveiled. It has been shown that only a fraction of NK clones expressing either KIR2DS1 or KIR2DS2 are able to lyse more efficiently target cells that express the HLA-C molecules recognized by their inhibitory counterparts (HLA-Cw4 and HLA-Cw3). KIR2DS4 is another member of the activating KIR group that has been cloned (BOTTINO et al. 1996). In the transmembrane and intracytoplasmic domains, it shows a high homology (99%) with the other two KIR2DS described above. By contrast, the two

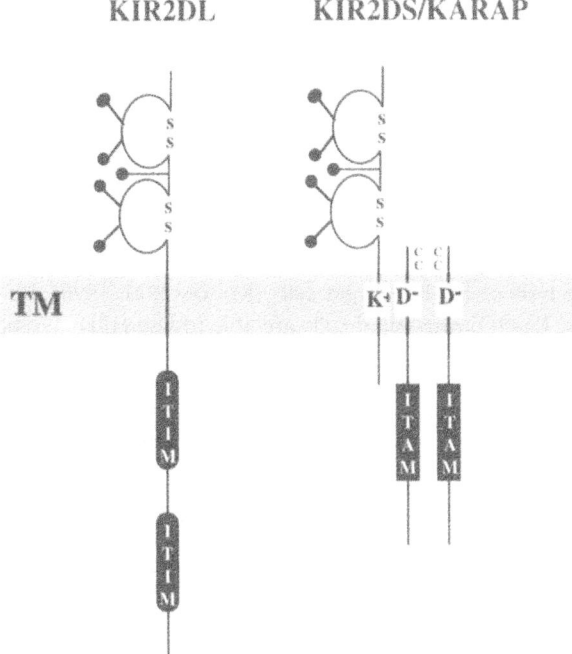

Fig. 1. Schematic representation of inhibitory and activating killer cell Ig-like receptors (KIR). On the *left* is an inhibitory KIR characterized by two Ig-like domains in the extracytoplasmic portion, a non-polar transmembrane region, and two ITIM motifs in the intracytoplasmic domain (KIR2DL). On the *right* is an activating KIR characterized by two Ig-like domains in the extracytoplasmic portion, a charged amino acid residue in the transmembrane portion, and a short intracytoplasmic region (KIR2DS). The activating KIR is associated with the ITAM-bearing polypeptide KARAP/DAP12

Ig-like domains are significantly different. Interestingly, it has been shown that a specific mAb for KIR2DS4 is able to induce selective proliferation of KIR2DS4[+] NK cells from unfractionated peripheral blood lymphocytes, suggesting that these molecules could play a role not only in the effector phases of NK cell activity (i.e., cytotoxicity), but could also regulate NK cell proliferation. The activating counterpart of KIR3DL1 has also been identified (DOHRING et al. 1996), but at present no functional data are available for this protein.

Activating counterparts of human lectin-like NKR have also been identified. The CD94 can associate with NKG2-C, a member of the NKG2 family characterized by approximately 95% amino acid identity with NKG2-A in the extracellular domain, a lysine in the transmembrane domain, and a mutation (Y to F) in the ITIM motifs (CANTONI et al. 1998). Triggering of CD94/NKG2-C heterodimers induces NK cell activation (HOUCHINS et al. 1997). Another member of the NKG2 family, NKG2-E, presents similar structural characteristics and might associate with CD94 to form another activating receptor (GLIENKE et al. 1998). Recently, it has been shown that both inhibitory and activating CD94/NKG2 heterodimers are able to bind HLA-E molecules (BRAUD et al. 1998; LEE et al. 1998). HLA-E, a nonclassical MHC class I molecule of limited sequence variability, requires binding of peptides derived from the signal sequence of some other MHC class I proteins for stable cell surface expression.

Two molecules belonging to the mouse Ly49 family have been described as potentially activating: Ly49D and Ly49H. Both receptors present an arginine in the transmembrane portion and mutations in the ITIM sequence. Ly49D extracytoplasmic domain shows homology with the Ly49 A inhibitory receptor, while the lectin-like domain of Ly49H is related to that of Ly49 C. Functional activation of mouse NK cell activity has been demonstrated only for Ly49D: mAb crosslinking of the receptor enhances the lysis of FcγR$^+$ target cells (MASON et al. 1996). Recently it has been shown that Ly49D transfected cells are able to lyse H2D^d target cells, but not target cells transfected with other H2 alleles, demonstrating that Ly49D is a receptor for H2D^d (NAKAMURA et al. 1999; GEORGE et al. 1999).

Both human- and mouse-activating NKR are characterized by the absence of transducing elements in their intracytoplasmic domain. Moreover, the presence of a positively charged amino acid residue in the transmembrane portion (either lysine or arginine) strongly suggested an interaction with specialized transducing polypeptides as for T cell receptor (TCR) with CD3γ, δ, ε, ζ and B cell receptor (BCR) with CD79α,β. A 12-kDa disulfide-linked tyrosine phosphorylated homodimer originally described as KARAP (KAR-associated polypeptide) associates with human activating KIR (OLCESE et al. 1997). Along the same line, a 16-kDa disulfide-linked tyrosine phosphorylated homodimer associated with Ly49D (MASON et al. 1998). These two molecules are identical and have been recently cloned and named KARAP/DAP12 (LANIER et al. 1998b; TOMASELLO et al. 1998). We will refer to KARAP/DAP12 for both human and mouse polypeptides. Human KARAP/DAP12 is a polypeptide containing a canonical immunoreceptor tyrosine-based activation motif (ITAM) close to CD3ζ and FcRγ molecules. Human KARAP/DAP12 gene is located on chromosome 19 close to the KIR family genes, whereas mouse KARAP/DAP12 gene is located on chromosome 7 close to genes coding for ITIM-bearing molecules. Structurally, human KARAP/DAP12 harbors a very short extracytoplasmic region (14 amino acid residues) with two cysteine residues and presents a negative charged amino acid (aspartic acid) in the transmembrane region and an ITAM sequence in the intracytoplasmic domain (LANIER et al. 1998b). Human KARAP/DAP12 is associated with both activating KIR (Fig. 1) and CD94/NKG2-C molecules (LANIER et al. 1998b; LANIER et al. 1998c). Upon NKR triggering, the tyrosines in the ITAM of KARAP/DAP12 are phosphorylated and recruit the SH2-containing protein tyrosine kinases ZAP-70 and Syk. Mouse KARAP/DAP12 has 73% amino acid identity with the human molecule, with conserved cysteines in the extracytoplasmic domain, an aspartic acid in the transmembrane portion, and an intracytoplasmic ITAM (TOMASELLO et al. 1998). The association between mouse KARAP/DAP12 and Ly49D/Ly49H has been reported (SMITH et al. 1998).

3 Hypothesis on the Functions of the Activating Receptors

Although the structure and the mechanisms by which activating NKR transduce stimulatory signals have been clarified, the biological function of these receptors is still an enigma. We will discuss the various hypotheses that have been proposed.

First, if both activating and inhibitory receptors recognize the same MHC class I molecule, they could play different roles in NK cell biology: the activating molecules could control proliferation during NK cell development and/or clonal amplification of mature NK cells, while the inhibitory receptors inhibit NK cell effector functions (i.e., cytotoxicity and cytokine production; COLONNA 1998).

Second, the affinity of the MHC ligand for the activating receptors could be inferior to that for the inhibitory counterparts. In this case, the activation balance would be in favor of the inhibitory molecules. In the case of viral infection, the presence of a viral peptide presented by the MHC class I molecules could enhance the affinity of the activating forms and trigger the killing of infected cells. Two studies utilized the same methodological approach to assess the direct binding of KIR to a panel of HLA class I transfectants (BIASSONI et al. 1997; WINTER et al. 1998). BIASSONI et al. produced soluble forms of KIR2DL1, specific for HLA-Cw4, and of its counterpart KIR2DS1. They showed that the binding to HLA-Cw4$^+$ cells was significantly weaker for KIR2DS1 than for the inhibitory form. Winter et al. could not observe binding of the KIR2DS1, KIR2DS2, and KIR2DS4 molecules tested to 11 different HLA class I$^+$ transfectants. These results might support either the hypothesis that the inhibitory and activating KIR molecules have different affinities for the same HLA class I alleles, or that they are able to recognize different peptides presented by HLA class I molecules.

Finally, if the activating NKR recognize a ligand different from MHC class I molecules, they should be considered as part of the numerous triggering receptors expressed by NK cells as CD2, NKR-P1, CD40L, β_2-integrins, NKp44, and NKp46 (RENARD et al. 1997; MORETTA et al. 1997). In this regard, it is possible that activating NKR could be the receptors for non-identified soluble factors involved in NK cell development and/or proliferation.

4 Concluding Remarks

Inhibitory and activating receptors belonging to the same molecular families have been conserved towards evolution. These receptors are not derived from alternative splicing, but are encoded by separate genes. At present we have only begun to understand the transducing pathways of inhibitory and activating receptors: inhibitory receptors are able to terminate the activation programs via the recruitment of proteine tyrosine phosphatases through their ITIM sequences; the activating receptors are coupled to ITAM-bearing polypeptides (KARAP/DAP12 or FcRγ), (Table 1), responsible for transducing activation signals. Many obscure

Table 1. Activating isoforms of immunoreceptor tyrosine-based inhibition motif (ITIM)-bearing molecules

Superfamily	Common name	Origin	Expression	Associated molecule	Ligand
IgSF	KIR2DS1	H	T, NK cells	KARAP/DAP-12	HLA-Cw4 (low affinity)
	KIR2DS2	H	T, NK cells	KARAP/DAP-12	HLA-Cw3 (low affinity)
	KIR2DS3	H	T, NK cells	KARAP/DAP-12	?
	KIR2DS4	H	T, NK cells	KARAP/DAP-12	?
	KIR3DS1	H	T, NK cells	KARAP/DAP-12	?
	ILT-1/LIR-7	H	Macrophages, dendritic cells, few NK cells	FcRγ	?
	ILT-7	H	?	FcRγ	?
	ILT-8	H	?	FcRγ	?
	LIR-6a	H	?	?	?
	LIR-6b	H	?	?	?
	PIR-A	M	Mast cells, B cells, macrophages, dendritic cells	FcRγ	?
	SIRP-β	H, M	Broad (hematopoietic and non-hematopoietic cells)	?	?
	FcγRIII	H, M	NK cells, neutrophils, macrophages, monocytes	FcRγ	aggregated IgG
	FcγRI	H, M	Neutrophils, macrophages, monocytes, dendritic cells	FcRγ	IgG
LSF	Ly49-D	M	T, NK cells	KARAP/DAP-12	H-2 D^d
	Ly49-H	M	T, NK cells	KARAP/DAP-12	?
	NKG2-C	H	T, NK cells	KARAP/DAP-12	HLA-E
	NKRP1-A	M	T, NK cells	FcRγ	?
	NKRP1-C	M	T, NK cells	?	?

These molecules belong either to the immunoglogulin superfamily (IgSF) or to the lectin C-type superfamily (LSF). Killer cell Ig-like receptors (KIR) are classified according to the nomenclature previously described by Long et al.
NK, natural killer; H, human; M, mouse.

points remain to be elucidated regarding the in vivo function of activating isoforms of ITIM-bearing molecules, such as the nature of the ligands, and the regulation of their cell surface expression.

Acknowledgments. This work is supported by institutional grants from INSERM, CNRS, and the Ministère de l'Enseignement Supérieur et de la Recherche, and specific grants from the Association pour la Recherche contre le Cancer (E.V.), and the Ligue Nationale contre le Cancer, axe "Immunologie des Tumeurs", (E.V.). A.C. is supported by a fellowship of the Training and Mobility of Researchers Programme.

References

Amigorena S, Bonnerot C, Drake JR, Choquet D, Hunziker W, Guillet JG, Webster P, Sautes C, Mellman I, Fridman WH (1992) Cytoplasmic domain heterogeneity and functions of IgG Fc receptors in B lymphocytes. Science 256:1808–12

Biassoni R, Cantoni C, Falco M, Verdiani S, Bottino C, Vitale M, Conte R, Poggi A, Moretta A, Moretta L (1996) The human leukocyte antigen (HLA)-C-specific "activatory" or "inhibitory" natural killer cell receptors display highly homologous extracellular domains, but differ in their transmembrane and intracytoplasmic portions. J Exp Med 183:645–50

Biassoni R, Pessino A, Malaspina A, Cantoni C, Bottino C, Sivori S, Moretta L, Moretta A (1997) Role of amino acid position 70 in the binding affinity of p50.1 and p58.1 receptors for HLA-Cw4 molecules. Eur J Immunol 27:3095–9

Borges L, Hsu ML, Fanger N, Kubin M, Cosman D (1997) A family of human lymphoid and myeloid Ig-like receptors, some of which bind to MHC class I molecules. J Immunol 159:5192–6

Bottino C, Sivori S, Vitale M, Cantoni C, Falco M, Pende D, Morelli L, Augugliaro R, Semenzato G, Biassoni R, Moretta L, Moretta A (1996) A novel surface molecule homologous to the p58/p50 family of receptors is selectively expressed on a subset of human natural killer cells and induces both triggering of cell functions and proliferation. Eur J Immunol 26:1816–24

Braud VM, Allan DS, O'Callaghan CA, Soderstrom K, D'Andrea A, Ogg GS, Lazetic S, Young NT, Bell JI, Phillips JH, Lanier LL, McMichael AJ (1998) HLA-E binds to natural killer cell receptors CD94/NKG2 A, B, and C. Nature 391:795–9

Burshtyn DN, Scharenberg AM, Wagtmann N, Rajagoplan S, Berrada K, Yi T, Kinet J-P, Long EO (1996) Recruitment of tyrosine phosphatase HCP by the killer cell inhibitory receptor. Immunity 4:77–85

Cantoni C, Biassoni R, Pende D, Sivori S, Accame L, Pareti L, Semenzato G, Moretta L, Moretta A, Bottino C (1998) The activating form of CD94 receptor complex: CD94 covalently associates with the Kp39 protein that represents the product of the NKG2-C gene. Eur J Immunol 28:327–38

Cella M, Dohring C, Samaridis J, Dessing M, Brockhaus M, Lanzavecchia A, Colonna M (1997) A novel inhibitory receptor (ILT3) expressed on monocytes, macrophages, and dendritic cells involved in antigen processing. J Exp Med 185:1743–51

Colonna M (1998) Unmasking the killer's accomplice. Nature 391:642–3

Daëron M, Latour S, Malbec O, Espinosa E, Pina P, Pasmans S, Fridman WH (1995) The same tyrosine-based inhibition motif, in the intracytoplasmic domain of FcγRIIB, regulates negatively BCR-, TCR-, and FcR-dependent cell activation. Immunity 3:635–46.

Dohring C, Samaridis J, Colonna M (1996) Alternatively spliced forms of human killer inhibitory receptors. Immunogenetics 44:227–30

Glienke J, Sobanov Y, Brostjan C, Steffens C, Nguyen C, Lehrach H, Hofer E, Francis F (1998) The genomic organization of NKG2 C, E, F, and D receptor genes in the human natural killer gene complex. Immunogenetics 48:163–173

Karre K, Welsh RM (1997) Viral decoy vetoes killer cell. Nature 386:446–7

Kharitonenkov A, Chen Z, Sures I, Wang H, Schilling J, Ullrich A (1997) A family of proteins that inhibit signaling through tyrosine kinase receptors. Nature 386:181–6

Kubagawa H, Burrows PD, Cooper MD (1997) A novel pair of immunoglobulin-like receptors expressed by B cells and myeloid cells. Proc Nat Acad Sci USA 94:5261–6

Lanier LL (1998a) NK cell receptors. Annu Rev Immunol 16:359–393

Lanier LL, Corliss B, Wu J, Leong C, Phillips JH (1998b) Immunoreceptor DAP12 bearing a tyrosine-based activation motif is involved in activating NK cells. Nature 391:703–7

Lanier LL, Corliss B, Wu J, Phillips JH (1998c) Association of DAP12 with activating CD94/NKG2 C NK cell receptors. Immunity 8:693–701

Lee N, Liano M, Carretero M, Ishitani A, Navarro F, Lopez-Botet M, Geraghty DE (1998) HLA-E is major ligand for the natural killer inhibitory receptor CD94/NKG2 A. Proc Natl Acad Sci USA 95:5199–5204

Long EO, Colonna M, Lanier LL (1996) Inhibitory MHC class I receptors on NK and T cells: a standard nomenclature. Immunol Today 17:100.

Mason LH, Anderson SK, Yokoyama WM, Smith HRC, RW-P, Ortaldo JR (1996) The Ly49D receptor activates murine natural killer cells. J Exp Med 184:2119–2128

Mason LH, Willette-Brown J, Anderson SK, Gosselin P, Shores EW, Love PE, Ortaldo JR, McWicar DW (1998) Characterization of an associated 16-kDa tyrosine phosphoprotein required for Ly49D signal transduction. J Immunol 160:4148–4151

Moretta A, Sivori S, Vitale M, Pende D, Morelli L, Augugliaro R, Bottino C, Moretta L (1995) Existence of both inhibitory (p58) and activatory (p50) receptors for HLA-C molecules in human natural killer cells. J Exp Med 182:875–84

Moretta A, Bottino C, Vitale M, Pende D, Biassoni R, Mingari MC, Moretta L (1996) Receptors for HLA class I molecules in human natural killer cells. Annu Rev Immunol 14:619–48

Nakamura MC, Linnemeyer PA, Niemi EC, Mason LH, Ortaldo JR, Ryan JC, Seaman WE (1999) Mouse Ly-49D recognizes H-2D^d and activates natural killer cell cytotoxicity. J Exp Med 189:493–500

Olcese L, Cambiaggi A, Semenzato G, Bottino C, Moretta A, Vivier E (1997) Human killer cell activatory receptors for MHC class I molecules are included in a multimeric complex expressed by natural killer cells. J Immunol 158:5083–6

Olcese L, Lang P, Vély F, Cambiaggi A, Marguet D, Blery M, Hippen KL, Biassoni R, Moretta A, Moretta L, Cambier JC, Vivier E (1996) Human and mouse killer-cell inhibitory receptors recruit PTP1 C and PTP1D protein tyrosine phosphatases. J Immunol 156:4531–4

Renard V, Cambiaggi A, Vély F, Blery M, Olcese L, Olivero S, Bouchet M, Vivier E (1997) Transduction of cytotoxic signals in natural killer cells: a general model of fine tuning between activatory and inhibitory pathways in lymphocytes. Immunol Rev 155:205–21

Samaridis J, Colonna M (1997) Cloning of novel immunoglobulin superfamily receptors expressed on human myeloid and lymphoid cells: structural evidence for new stimulatory and inhibitory pathways. Eur J Immunol 27:660–5

Sato S, Tuscano JM, Inaoki M, Tedder TF (1998) CD22 negatively and positively regulates signal transduction through the B lymphocyte antigen receptor. Semin Immunol 10:287–297

Smith KM, Wu J, Bakker ABH, Phillips JH, Lanier LL (1998) Ly49D and Ly49H associate with mouse DAP12 and form-activating receptors. J Immunol 161:7–10

Takei F, Brennan J, Mager DL (1997) The Ly-49 family: genes proteins and recognition of class I MHC. Immunol Rew 155:67

Tomasello E, Olcese L, Vély F, Geourgeon C, Bléry M, Moqrich A, Gautheret D, Djabali M, Mattei MG, Vivier E (1998) Gene structure, expression pattern and biological activity of mouse KAR-associated protein KARAP/DAP12. J Biol Chem. 273:34115

Vance RE, Kraft JR, Altman JD, Jensen PE, Raulet DH (1998) Mouse CD94/NKG2 A is a natural killer cell receptor for the nonclassical major histocompatibility complex (MHC) class I molecule Qa-1^b. J Exp Med 188:1841–1848

Vély F, Vivier E (1997) Conservation of structural features reveals the existence of a large family of inhibitory cell surface receptors and noninhibitory/activatory counterparts. J Immunol 159:2075–7

Vivier E, Daëron M (1997) Immunoreceptor tyrosine-based inhibition motifs. Immunol Today 18:286–91

Wagtmann N, Rojo S, Eichler E, Mohrenweiser H, Long EO (1997) A new human gene complex encoding the killer cell inhibitory receptors and related monocyte/macrophage receptors. Curr Biol 7:615–618

Winter CC, Gumperz JE, Parham P, Long EO, Wagtmann N (1998) Direct binding and functional transfer of NK cell inhibitory receptors reveal novel patterns of HLA-C allotype recognition. Curr Biol 8:1009–1017

Wu Y, Nadler MJS, Brennan LA, Gish GD, Timms JF, Fusaki N, Jongstra-Bilen J, Tada N, Pawson T, Wither J, Neel BG, Hozumi N (1998) The B-cell transmembrane protein CD72 binds to and is an in vivo substrate of the protein tyrosine phosphatase SHP-1. J Immunol 161:571–577

Subject Index

Current Topics in Microbiology and Immunology

Volumes published since 1989 (and still available)

Vol. 223: **Tracy, S.; Chapman, N. M.; Mahy, B. W. J. (Eds.):** The Coxsackie B Viruses. 1997. 37 figs. VIII, 336 pp. ISBN 3-540-62390-6

Vol. 224: **Potter, Michael; Melchers, Fritz (Eds.):** C-Myc in B-Cell Neoplasia. 1997. 94 figs. XII, 291 pp. ISBN 3-540-62892-4

Vol. 225: **Vogt, Peter K.; Mahan, Michael J. (Eds.):** Bacterial Infection: Close Encounters at the Host Pathogen Interface. 1998. 15 figs. IX, 169 pp. ISBN 3-540-63260-3

Vol. 226: **Koprowski, Hilary; Weiner, David B. (Eds.):** DNA Vaccination/Genetic Vaccination. 1998. 31 figs. XVIII, 198 pp. ISBN 3-540-63392-8

Vol. 227: **Vogt, Peter K.; Reed, Steven I. (Eds.):** Cyclin Dependent Kinase (CDK) Inhibitors. 1998. 15 figs. XII, 169 pp. ISBN 3-540-63429-0

Vol. 228: **Pawson, Anthony I. (Ed.):** Protein Modules in Signal Transduction. 1998. 42 figs. IX, 368 pp. ISBN 3-540-63396-0

Vol. 229: **Kelsoe, Garnett; Flajnik, Martin (Eds.):** Somatic Diversification of Immune Responses. 1998. 38 figs. IX, 221 pp. ISBN 3-540-63608-0

Vol. 230: **Kärre, Klas; Colonna, Marco (Eds.):** Specificity, Function, and Development of NK Cells. 1998. 22 figs. IX, 248 pp. ISBN 3-540-63941-1

Vol. 231: **Holzmann, Bernhard; Wagner, Hermann (Eds.):** Leukocyte Integrins in the Immune System and Malignant Disease. 1998. 40 figs. XIII, 189 pp. ISBN 3-540-63609-9

Vol. 232: **Whitton, J. Lindsay (Ed.):** Antigen Presentation. 1998. 11 figs. IX, 244 pp. ISBN 3-540-63813-X

Vol. 233/I: **Tyler, Kenneth L.; Oldstone, Michael B. A. (Eds.):** Reoviruses I. 1998. 29 figs. XVIII, 223 pp. ISBN 3-540-63946-2

Vol. 233/II: **Tyler, Kenneth L.; Oldstone, Michael B. A. (Eds.):** Reoviruses II. 1998. 45 figs. XVI, 187 pp. ISBN 3-540-63947-0

Vol. 234: **Frankel, Arthur E. (Ed.):** Clinical Applications of Immunotoxins. 1999. 16 figs. IX, 122 pp. ISBN 3-540-64097-5

Vol. 235: **Klenk, Hans-Dieter (Ed.):** Marburg and Ebola Viruses. 1999. 34 figs. XI, 225 pp. ISBN 3-540-64729-5

Vol. 236: **Kraehenbuhl, Jean-Pierre; Neutra, Marian R. (Eds.):** Defense of Mucosal Surfaces: Pathogenesis, Immunity and Vaccines. 1999. 30 figs. IX, 296 pp. ISBN 3-540-64730-9

Vol. 237: **Claesson-Welsh, Lena (Ed.):** Vascular Growth Factors and Angiogenesis. 1999. 36 figs. X, 189 pp. ISBN 3-540-64731-7

Vol. 238: **Coffman, Robert L.; Romagnani, Sergio (Eds.):** Redirection of Th1 and Th2 Responses. 1999. 6 figs. IX, 148 pp. ISBN 3-540-65048-2

Vol. 239: **Vogt, Peter K.; Jackson, Andrew O. (Eds.):** Satellites and Defective Viral RNAs. 1999. 39 figs. XVI, 179 pp. ISBN 3-540-65049-0

Vol. 240: **Hammond, John; McGarvey, Peter; Yusibov, Vidadi (Eds.):** Plant Biotechnology. 1999. 12 figs. XII, 196 pp. ISBN 3-540-65104-7

Vol. 241: **Westblom, Tore U.; Czinn, Steven J.; Nedrud, John G. (Eds.):** Gastroduodenal Disease and Helicobacter pylori. 1999. 35 figs. XI, 313 pp. ISBN 3-540-65084-9

Vol. 242: **Hagedorn, Curt H.; Rice, Charles M. (Eds.):** The Hepatitis C Viruses. 1999. 47 figs. approx. IX, 380 pp. ISBN 3-540-65358-9

Vol. 243: **Famulok, Michael; Winnacker, Ernst-L.; Wong, Chi-Huey (Eds.):** Combinatorial Chemistry in Biology. 1999. 48 figs. IX, 189 pp. ISBN 3-540-65704-5

Springer
and the
environment

At Springer we firmly believe that an international science publisher has a special obligation to the environment, and our corporate policies consistently reflect this conviction.

We also expect our business partners – paper mills, printers, packaging manufacturers, etc. – to commit themselves to using materials and production processes that do not harm the environment. The paper in this book is made from low- or no-chlorine pulp and is acid free, in conformance with international standards for paper permanency.

 Springer

The manufacturer's authorised representative in the EU is Springer
Nature Customer Service Centre GmbH, Europaplatz 3, 69115 Heidelberg,
Germany. If you have any concerns regarding our products, please
contact ProductSafety@springernature.com

Printed and bound by CPI Group (UK) Ltd, Croydon, CR0 4YY

28/04/2026

02098488-0001